MISSILE AND SPACE PROJECTS GUIDE
1962

MISSILE AND SPACE PROJECTS GUIDE 1962

Horace Jacobs

and

Eunice Engelke Whitney

Springer Science+Business Media, LLC

ISBN 978-1-4899-6211-9   ISBN 978-1-4899-6427-4 (eBook)
DOI 10.1007/978-1-4899-6427-4

Softcover reprint of the hardcover 1st edition 1962

DEDICATION

Dedicated to our Earthling children,
World Citizens of the Space Age.

PREFACE

The Missile and Space Projects Guide has been prepared to provide aerospace scientists, engineers, market analysts, planners, and other specialists with basic information on the numerous existing missile and space projects. It is the first book of its kind in that it assembles in one volume reference information on nearly all projects, planned or actual, including concepts, studies, or hardware, that have been mentioned in the open literature. A number of missile and space guides, dictionaries, glossaries, or handbooks have been published, but none is centered about the "project" in its broadest sense. Some of these guides furnish detailed information on a limited number of R&D projects but ignore the hundreds of studies, proposals, concepts, and programs which complete the picture.

If we consider missile and space projects as trees in a forest, it can be said that the mighty oaks and many of the larger trees have been covered in the literature, but the young growth, from which the big trees of the future must develop, has not been treated extensively in existing guides. It is the aim of this volume to provide reference information on all missile and space projects, whether small or great, domestic or foreign, conceptual or material.

Apart from the names of actual missile and space programs, this guide includes numerical designations of such projects and of pieces of associated equipment, as well as the names or abbreviations of many aerospace organizations. Nearly all known missile and space project names and project designations have been identified and cross-referenced. All are not included because new ones are coming into existence daily. However, more projects than ever before assembled are listed and identified in a systematic manner.

Obviously, the present guide cannot provide a quantity of detailed information on any one project, but it does provide sufficient data to orient the reader, classify the project, and relate it to an organization or sponsoring agency. Where possible, budgetary and cost information, principal subcontractors, status of the project, and some descriptive data are given. Although every effort has been made to check the validity of the data presented, the authors cannot guarantee the accuracy of every item. This derives from the fact that missile and space programs are under constant revision and that only unclassified sources have been used in this compilation. For the present guide to be useful, it must be current; it could not be so if the authors had not struck a suitable balance between time and accuracy.

The Authors

ABBREVIATIONS

The following abbreviations were used in the preparation of this Guide. Many other abbreviations are defined in the body of the work.

ABL	Allegheny Ballistics Laboratory
ACF	ACF Industries, Inc.
AMF	American Machine & Foundry Co.
APL	Applied Physics Laboratory (at Johns Hopkins University)
APU	Auxiliary Power Unit
ATT	American Telephone & Telegraph Co.
BTZ	Bureau Technique Zborowski
CEP	Circular Error Probability
CRL	Cambridge Research Laboratory
DEFA	Direction des Etudes et Fabrications d'Armement
ECM	Electronic Countermeasure
FTL	Federal Telecommunications Laboratory
FY	Fiscal Year
G.D.	General Dynamics Corp.
G.E.	General Electric Co.
G.M.	General Motors Corp.
HE	High Explosive
HF	High Frequency
HRB	HRB-Singer, Inc.
IBM	International Business Machines Corp.
IGY	International Geophysical Year
IR	Infrared
ITT	International Telephone & Telegraph Corp.
JATO	Jet-Assisted Take-Off
JHU	Johns Hopkins University
HPL	Jet Propulsion Laboratory
LHY	Liquid Hydrogen
LOX	Liquid Oxygen
NAA	North American Aviation, Inc.
R&D	Research and Development
RCA	Radio Corporation of America
RDT&E	Research, Development, Test, and Engineering
SRI	Stanford Research Institute (Menlo Park, Calif.)
STL	Space Technology Laboratories, Inc.
TRW	Thompson Ramo Wooldridge, Inc.
UHF	Ultrahigh Frequency

VHF	Very High Frequency
VLF	Very Low Frequency
VTOL	Vertical Take-Off and Landing
WECO	Western Electric Corp.

MISSILE AND SPACE PROJECTS GUIDE
1962

124-E

Special Army configuration of the advanced FIREBEE. Ryan Aeronautical Co.

480-1 (A-4)

Air Command global communications network.

1234, 12345

Military designation for Air Force basic HETS 609A. Model 12345 has larger fourth stage.

2356

See ARGO D-8 (JOURNEYMAN).

A

A1-F

Also A1-X, A1-XE7, A1-XE8. Military designations for Navy POLARIS test vehicle.

A3H (Argentina – Air Force)

Air-to-surface missile. Length, about 3 ft. Weight, 48 lb. Solid propellant. Range, 3.8 miles. Free guidance. High-explosive warhead. Developed under Argentine Army contract for the Air Force by Instituto de Investigaciones Técnicas.

A-4

See 480-1.

A-9 (Germany)

Early German test vehicle, considered as an alternative design to the V-2, but never produced. Similar specifications to those of the V-2, but with a 75,000-lb thrust. Later version was to have been piloted. Also was to have been upper stage of the A-9/A-10 project (AMERIKA RAKETE project).

A-10 (Germany)

Early German test vehicle. Length, 65 ft. Diameter, 162 in. Weight, loaded, 175,000 lb. With the A-9 as upper stage, formed an intercontinental ballistic missile. However, the vehicle never passed the design stage.

AA-20, AA-30 (France – Air Force)
Air-to-air and air-to-surface missiles. Length, 8.5 ft. Launch
weight (A/A version), 295 lb; (A/S version), between 295 and
375 lb. Solid propellant. Range, 2.5 miles. Supersonic. High-
explosive warhead. Deployed with Air Force and Navy units.
Adaptation, ACAM 5301, under development for Army; follow-
on AA-30 under development for Air Force; and ACAM 5103
adopted for NATO. Nord Aviation.

AAM
General designation for an air-to-air missile.

AAMM (Air Force)
Anti-antimissile missile. Study by University of Michigan.

AAM-N-2
See SPARROW 1.

AAM-N-3
See SPARROW 2.

AAM-N-6
See SPARROW 3.

AAM-N-7
See SIDEWINDER.

ABLE (Air Force NASA)
The term ABLE was used to describe a booster configuration
consisting of a modified VANGUARD vehicle for the upper stages
boosted by an ATLAS or a THOR. The ABLE consists of an
upper solid stage and lower liquid stages. Operational, phasing
out 1961. See ATLAS ABLE, THOR ABLE, VANGUARD. Space
Technology Laboratories/G. D.-Astronautics.

ABLE (Army)
Lightweight missile for direct support to battle group. Feasibil-
ity studies to Armour, Cornell Aeronautical Laboratory, G. E.,
Minneapolis-Honeywell, Martin, and Douglas.

ABLE (Army)
Portable, light gyroscopic direction-finder. Autonetics.

ABLE (Navy)
See WEAPON ALFA.

ABLE III
    See EXPLORER VI.

ABLESTAR I, II (Army/Navy/Air Force)
    An improved ABLE configuration for use as second stage for
    TRANSIT and COURIER launchings. First stage will be THOR,
    ATLAS, or TITAN. ABLESTAR I first used April 1960. ABLE-
    STAR II scheduled for future space missions. Now in R & D.
    Aerojet—General.

ABMA
    Army Ballistic Missile Agency. Renamed ARGMA in 1960, then
    reorganized in January 1962. All missile programs now under
    Army Ordnance Missile Command in 13 weapon system project
    offices.

ACAM 5301 (France—Army)
    Surface-to-air missile intended for use by combat troops against
    low-flying aircraft. Solid propellant. Range, nearly 10 nm.
    Speed, about Mach 2. Radar guidance. Nord Aviation.

ACDC
    Army Combat Development Command. Purpose: to perform
    the combat development functions formerly assigned to CONARC,
    Technical and Administrative Services, and other agencies.

ACE
    Aerospace Control Environment. Term used by Air Force,
    especially at Hanscom Field, for controlling environment in
    space.

ACEC (Belgium)
    Ateliers de Constructions Electriques de Charleroi. Belgian
    manufacturing firm. See also SETEL.

ACES
    Automatic Controls Evaluation Simulator. Capable of duplica-
    tion of control of a vehicle in space as nearly as possible. An
    aluminum tubing platform 12 in. in length, weighing about
    1350 lb. The base is a 4.5-in. steel ball supported on a column
    of nitrogen gas at 120-psi pressure. Platform has a seat, side-
    mounted controller, and foot pedals for manned operation. Ling-
    Temco Vought, Dallas, Tex.

ACHIEVER (Air Force)
Inertial guidance system used in TITAN booster. Has capability of "remembering" simultaneously the routes to several targets. Weight, 90 lb. System used also with MACE, THOR, G. M.—AC Electronics. Div.

ACRE (Navy)
Automatic Checkout and Readiness Equipment for missile launching. Combined with POLARIS as subsystem for submarine use. Lockheed.

ACS (Army)
Alaska Communication System. A conventional open-wire and buried-cable telephone system. Capital investment about $45.5 million.

ACV
Air Cushion Vehicle. Operation height is 3 in. off road. Bell Aerosystems.

ADAM (Army)
A proposed means of getting man into space for a few minutes by rocketing an occupant to about 150 miles in a REDSTONE nose section which would have been parachute-recoverable. Initially rejected by ARPA. However, the MERCURY program has done precisely this.

ADAM
See FRIDA (Sweden).

ADC
Air Defense Command.

ADOS
Astronautical Defensive-Offensive System. A system proposed by Avco in 1959 to the Air Force to protect space vehicles from attack or piracy in space. The one-man vehicle consists of a spherical nuclear reactor unit at the end of a long cylinder. The sphere is surrounded by a nuclear accelerator ring. Since the ring is movable on two axes, the pilot can fire nuclear weapons in any direction.

ADVD – 58
See TRAILBLAZER.

ADVENT (Army)
A 24-hr communication satellite program under the over-all direction of Aerospace Corp. for the Army Signal Corps. A real-time repeater satellite weighing about 1000 lb for launch into 22,300 mile orbit around the equator. Three 6-hr orbit test shots using ATLAS-AGENA boosters are planned for the 1961-1962 time period and the firing of ADVENT satellites in the 1962-1964 time period. Incorporates former projects STEER, TACKLE, and DECREE. ATLAS-CENTAUR booster. Part of NOTUS program. Prime is Army Signal Corps. Bendix is prime for the communications package, and G. E. for vehicle design.

AEC
U. S. Atomic Energy Commission.

AEDC (Air Force)
Arnold Engineering Development Center, Tullahoma, Tenn.

AEOLUS
Weather-sounding rocket. Rocketdyne-sponsored program. Designed to reach altitudes equal to ARCAS.

AERIE
Air defense system proposed by Bendix. Would use modified EC-135 transport aircraft equipped with search and acquisition radar and would have a battle control center capable of directing more than 24 EAGLES to target.

AEROBEE (Army/Navy/Air Force/NASA)
An Aerojet-General family of sounding rockets used by all three services and NASA for research in the upper atmosphere. Research projects include studies of the behavior of liquid hydrogen in a zero-gravity environment, testing an attitude control system of interest to the Orbiting Astronomical Observatory (OAO), study of solar radiation, stellar observations, celestial ultra-violet radiation measurements, and collection of data on micro-meteorites. Some designations for these vehicles are: Navy - RTV-N-10; RTV-N-10b; RTV-N-8al; AJ10-24; AJ11-18; AGVL-0113C; AGVL-0113F; RVN-13A,B; XRN-N-13. Air Force - RTV-A-1, 1b, 1a; AJ10-25; AJ11-6; AJ10-34; AJ11-20; AJ11-21; AGVL-0113I; AGVL-0113H. Army - XASR-SC-1, 2. Cost is approximately $20,000 to $40,000 per rocket. See specific rockets in AEROBEE series.

AEROBEE 75 (Army)
Also known as AEROBEE HAWK. Aerojet-General dual-thrust, solid-propellant rocket used principally by the Army. Can carry about 90 lb to 75-mile altitude.

AEROBEE 100 (NASA)
Also called AEROBEE JR. Developed in 1958 as a medium-performance rocket for lower-atmosphere use up to about 100 miles. It consists of a scaled-down AEROBEE engine plus a solid booster. Cost is approximately $20,000 each. Can carry about 50 lb to 90 miles or 100 lb to 70 miles.

AEROBEE 105A (NASA)
A two-stage Aerojet-General sounding rocket designed to test an attitude control system. Of special interest to the Orbiting Astronomical Observatory (OAO), a NASA program. Planned to boost 195 lb to 134 miles at a maximum velocity of 6100 ft/sec.

AEROBEE 150 (NASA)
Formerly called AEROBEE-HI Aerojet-General fixed-fin, second-stage sounding rocket. Solid propellant. Capable of sending 160 lb to about 150-mile altitude. Thrust, about 4100 lb. AEROBEE 150A used to test behavior of liquid hydrogen in a zero gravity environment for NASA. A 300-lb package is boosted to about 100 miles for this purpose.

AEROBEE 300
Also called SPAEROBEE. An Aerojet-General sounding rocket consisting of an AEROBEE 150 first stage and a solid-propellant SPARROW I second stage. Capable of lofting 50 lb to 300-mile altitude.

AEROBEE HAWK
See AEROBEE 75.

AEROBEE-HI
See AEROBEE 150.

AEROBEE JR.
See AEROBEE 100.

AEROGARD
Plastic foam packaging system for enclosing delicate components. Aerojet-General.

AEROPAK (Army)
Feasibility study by Aeroject-General for a rocket pack to transport a soldier over difficult terrain. Similar to Thiokol proposal for a rocket-powered jump belt.

AEROS (NASA)
Third-generation weather satellite currently planned to be launched by ATLAS-AGENA B into synchronous earth orbit at 22,300-mile altitude. Payload not yet fully defined. Preliminary system studies and development of variable focus-type lens set for 1962. Six-year plan for an operational system to begin in 1964 has been approved, using four spacecraft. Thereafter, two AEROS satellites to be launched per year. Major problems in AEROS system considered to be a reliable camera system, and attitude and orbit correction. AEROS to supplement the polar-orbiting NIMBUS vehicles. Cost: $60 million annually, of which approximately $49 million to be used for four spacecraft and four boosters.

AEROSCAR (Navy)
An experimental solid-propellant missile for use in the HYDRA program which calls for launch of space payloads from the surface of the ocean.

AEROSCORE C
Optical electronic system for measuring trajectories of missiles. Used in drones. Aerojet-General.

AEROSOUND (Air Force)
A variation of the Aerojet-General AEROBEE. Launched vertically from an airplane. A one-stage version can carry 20 lb to about 55 miles or 40 lb to about 45 miles. A two-stage version can carry such payloads to over 100 miles. Army and Navy also reported interested.

AEROSPACE PLANE (Air Force)
Also designated ASP and SPACE PLANE. Winged plane designed to fly both in atmosphere and in space. Gathers its own oxygen as it flies through the atmosphere. Flies to 300-nm orbit and returns for re-entry, as in DYNA-SOAR, X-15 fashion. Proposed molybdenum construction. Turbine, ramjet, LXO propulsion. Revealed at AFA convention, 1960. In design study phase. Design studies made by Lockheed, Douglas, Republic, Corvair, NAA, Martin. Air Force estimates craft can be developed by about 1970.

AEROWOLF
Aerojet-General proposal for an air-to-air, solid-propellant missile designed to seek and destroy enemy fighters and bombers. Missile system rejected in 1952.

AESOP (Navy)
Artificial Earth Satellite Observation Program. Detects silent satellites automatically by infrared preselectors. Navy version of BAMBI. In R&D. Contract for design and manufacture of infrared preselectors to Hermes Electronic.

AF-AA10.02
Military designation for Air Force SPAEROBEE.

AFA
Air Force Association.

AFASD
Also ASD. Air Force Aeronautical Systems Division located at Wright-Patterson AFB. Combines the Aeronautical Systems Center of AMC and the Wright Air Development Division of ARDC. Responsible for such programs as SKYBOLT, DYNA-SOAR, and the C-141 jet transport. See AFSC.

AFB
Air Force Base.

AFBMD
Also BMD. Air Force Ballistic Missile Division, Inglewood, Calif. Now changed to AFBSD.

AFBSD
Also BSD. Air Force Ballistic Systems Division, Inglewood, Calif., a division of the Air Force Systems Command. Formerly AFBMD. Composed of elements of the earlier ARDC Ballistic Missile Division, the AMC Ballistic Missile Center, and the Ballistic Missile Office of the Army Corps of Engineers which is in charge of ballistic missile base construction. See AFSC.

AFESD
Also ESD. Air Force Electronic Systems Division, Hanscom AFB, Boston, Mass. Composed of the former Command and Control Development Division of ARDC and the former Electronics Systems Center of AMC. AFESD is the only division with a center (Rome Air Development Center, Griffiss AFB, N. Y.) specifically assigned to its control.

AFFTC
Air Force Flight Test Center, Edwards AFB, Calif.

AFLC
Air Force Logistics Command. Formerly Air Materiel Command (AMC).

AFMDC
Air Force Missile Development Center, Holloman AFB, N. M.

AFMTC
Air Force Missile Test Center, Patrick AFB, Fla.

AFOAR
Also OAR. Air Force Office of Aerospace Research. Responsible directly to the Air Force Chief of Staff. In charge of basic research.

AFOSR
Also OSR. Air Force Office of Scientific Research.

AFSC
Air Force Systems Command. Formerly ARDC, Air Research and Development Command. All activities concerning development and acquisition of aircraft and missile systems were consolidated in this command under Gen. Bernard A. Schriever, as of 1 July 1961. Four divisions of AFSC: Ballistic Systems Division (BSD), Aeronautical Systems Division (ASD), Space Systems Division (SSD), and the Electronics Systems Division (ESD).

AFSSD
Also SSD. Air Force Space Systems Division of AFSC, composed of elements of the former ARDC Ballistic Missile Division and AMC Ballistic Missile Center. Located in Inglewood, Calif. Responsible for military space programs assigned to the USAF and other programs in support of NASA, Navy, and Army. See AFSC.

AFSWC
Air Force Special Weapons Center, Kirtland AFB, N. M., at which atomic weapons and systems components are developed and tested, and personnel hazards associated with them evaluated.

AGARD
Advisory Group for Aeronautical Research and Development, North Atlantic Treaty Organization.

AGATE (France)
Missile capable of attaining a 40-mile altitude.

AGE
Aerospace Ground Equipment. Air Force term for Ground Support Equipment (GSE).

AGENA B, C (Air Force/NASA)
Vehicle used as second stage with ATLAS and THOR in boosting large payloads into earth and moon orbits. Used with THOR in DISCOVERER program. Height, 25 ft. Weight, 15,550 lb with full propellant. Orbital weight, 2100 lb. Study being made of AGENA C (used with ATLAS). To be capable of lofting 2200 lb into 3500-mile orbit or 1000 lb into 22,300-mile orbit. Lockheed Missiles and Space, prime; Bell, propulsion.

AGILE (Navy)
Air-to-air missile. Classified. Bendix/Grumman.

AGREE (Air Force)
Advisory Group on Reliability of Electronic Equipment. A group formed to establish electronic standards to increase reliability.

AGVL-0113C
Also AGVL-0113H. Military designations for Air Force models AEROBEE-HI.

AGVL-0113I
Also AGVL-0113H. Military designations for Air Force models of AEROBEE-HI.

AICBM
Anti-InterContinental Ballistic Missile.

AIGLE (France)
Single-stage high-altitude sounding rocket. 1320-lb booster charge. Rocket will loft 220-lb payload to 120-mile altitude or 660-lb payload to 55-mile altitude. Solid propellant. One of a family of low-cost rockets being developed by Sud-Aviation. See BELIER, CENTAURE, DRAGON, ERIDAN, PEGASE.

AIRBM
Anti-Intermediate Range Ballistic Missile.

AIRONE (Italy - Army)
Surface-to-surface tactical missile. Solid booster. Range, six miles. Free guidance. Operational. Polverificio Giovanni Stacchini.

AJ10-24
Also AJ10-27, AJ11-18. Military designations for Navy models of AEROBEE research rocket.

AJ10-25
Also AJ10-34, AJ11-6, AJ11-20, AJ11-21. Military designations for Air Force models of AEROBEE research rocket.

AJ-10-115
Supersonic rocket sled developed by Aerojet-General.

ALARM (Army)
Automatic Light Aircraft Readiness Monitor. Concept being studied by Bendix of applying electronic checkout techniques now used in missile launching to determine whether Army aircraft are safe for flight.

ALBATROSS (Navy)
Proposed reconnaissance satellite for ocean areas.

ALBM
Air-Launched Ballistic Missile. See SKYBOLT.

ALDEBARAN
Martin proposal, 1960, for a 50,000 ton, winged space ship. Capable of lofting a 60-million-lb payload into a minimum earth orbit, or a 45-million-lb payload to the moon. Nuclear.

ALERT
A program to alert Americans to the danger to the Western nations of aggression and infiltration.

ALFA
See WEAPON ALFA.

ALFA TRAINER (NASA)
Air Lubricated Free Attitude trainer. Used in training MER-

CURY astronauts. Consists of a contoured couch mounted on an air bearing which is essentially frictionless. Trainer can be stabilized and controlled about all three axes by a MERCURY hand-controller actuating compressed-air jets.

ALGOL

First stage of SCOUT booster. Largest solid rocket launched in the U. S. by 1960. Length, 30 ft. Weight, 23,600 lb. Develops 115,000-lb thrust. Controlled in flight by jet vanes. Fin-stabilized. Outstanding reliability. See SCOUT. Aerojet-General.

ALOUETTE (Canada/NASA)

Sweep-frequency topside ionosphere sounder satellite for Canada. Also called TOPSIDE SOUNDER. Joint U. S.-Canada operation. Weight, approximately 275 lb. Cost, $3.25 million for four satellites. ALOUETTE I to be lofted by THOR-DELTA booster in early 1962. DeHavilland Aircraft of Canada.

ALOO (AEC)

ALbuquerque Operations Office of AEC. Provides technical direction of AEC work done by Los Alamos Scientific Laboratory on projects such as ROVER.

ALPHA

See WEAPON ALFA.

ALPHA ABC (Air Force)

Advanced Biomedical Capsule. 50-lb satellite capsule to take chimp through lower Van Allen belt. Phase Two called BRAVO; Phase Three, COCA. Lockheed Missiles and Space study.

ALPHA DRACO (Air Force)

Air-to-surface experimental missile entered by McDonnell Aircraft in the BOLD ORION (later SKYBOLT) competition. Solid propellant. Mission: to demonstrate the feasibility of air-launched ballistic missiles.

ALR

See GIMLET.

ALRI

Airborne Long-Range Input. Seward extension of SAGE system by radar station installed in an RC-121 plane. ALRI airborne tests completed in mid-1961. See SLRI.

ALSOR (NASA)
Air-Launched SOunding Rocket. Boosted by VIPER 1C rocket carried by F-104 and launched at about 40,000 ft, just after each X-15 flight. Rocket releases 12-ft balloon to determine air density, temperature, and winds and evaluate flight data beyond 90,000-ft altitude.

ALSPANO
German-Spanish dictionary compiled by machine translation. University of Mexico.

ALTAIR (NASA/Navy)
Plastic rocket used as fourth stage of SCOUT in the Air Force HETS high-altitude re-entry tests. Also used by NASA and Navy. Length, 6 ft. Weight, 500 lb. Develops over 3000 lb thrust. More than 90% solid propellant by weight. Hercules Powder.

ALUMINAUT
All-aluminum submarine designed for deep-ocean research. Leased by Woods Hole Oceanographic Institution, under contract to the Navy. To operate at 15,000 ft depths. Length, 50 ft. Launch date, 1963. Cost, $2 million. Designed by Reynolds Metals Co. To be built by G.D.—Electric Boat.

AMC
Air Materiel Command. Now redesignated the Air Force Logistics Command (AFLC).

AMERIKA RAKETE (Germany)
See A-9.

AMES RESEARCH CENTER (NASA)
NASA center located at Moffett Field, Calif., with the principal assignments in the areas of re-entry and human factors.

AMM
AntiMissile Missile.

AMR
Atlantic Missile Range.

AMSAM
AntiMissile Surface-to-Air Missile.

AMY (Air Force)
No details available on this research and test vehicle.

AN/ANQ-15
Weather reconnaissance system project. Bendix/Boeing.

AN/GJQ-9
Automatic checkout system for SKYBOLT and HOUND DOG. Bendix.

ANNA (Navy/Air Force/Army/NASA)
Formerly FIREFLY. Geodetic satellite to measure long distances on earth with accuracy to a few yards. Weight, 50 to 100 lb. Navy supplies TRANSIT-type Doppler system. Air Force supplies flashing-light system. Army supplies the positional device. NASA's role that of an observer, more or less. THOR-AGENA B booster. Launch, probably 1961 or early 1962.

ANP
Aircraft Nuclear Propulsion.

ANP-NUCLEAR ROCKET (Air Force)
Proposed series of short-lived satellites swooping down from 300-nm orbit to 100 mile altitude for detailed reconnaissance with optical TV or radar. Might carry three men. Pratt & Whitney. Canceled.

ANPO (AEC/Air Force)
Aircraft Nuclear Propulsion Office, Germantown, Md. May be abolished.

AN/SQS-26
Search sonar system. Means of detecting submarines at (classified)-mile range.

ANTARES
Winged space vehicle with captive acceleration system, turbojet sled, to soft-land 40-million-lb payload on the moon. Vehicle to weigh 20 million lb. Proposed by Martin, 1960.

ANTARES (France)
Four-stage sounding rocket evolved from the ONERA 56/39-22D. Solid propellant. An IRBM that provides over 13,600 lb thrust. Office National d'Etudes et de Recherches Aéronautiques (ONERA).

ANTARES (NASA)
Third stage of SCOUT. An improved VANGUARD third stage. Length, 10 ft. Weight, 2600 lb. Thrust, more than 13,600 lb.

More than 90% propellant by weight. See SCOUT. Hercules
Powder.

AN/TPD-2 (Army Signal Corps)
Experimental, portable radar system developed by Strand Engineering.

AN/USD-2
See SD-2.

AN/USD-4.
See SWALLOW.

AN/USD-5
See OSPREY.

AOMC
Army Ordnance Missile Command. See ABMA.

AP-2
Advanced version of POLARIS. Lockheed.

APACHE
Surface-to-air sounding rocket. Capable of carrying 35-lb payload to almost 40-mile altitude. Solid propellant. Patterned after
CAJUN. Thiokol/New Mexico State University.

APATS
Automatic Programmer And Test System. Used for environmental test chamber for satellites. Built for Lockheed by RCA.

APGC
Air Proving Ground Center, Eglin AFB, Fla.

APOLLO
Follow-on project to MERCURY. APOLLO capsule to weigh between 100,000 and 150,000 lb. Height, 15 to 20 ft. Length, about
60 ft. Command and service modules to be 12 to 15 ft in length
and in diameter. Lunar-landing and orbiting-laboratory modules
to be about 30 ft in length. To be launched by SATURN C-1 for
orbital flights, SATURN C-5 for lunar rendezvous-type missions,
or NOVA, if direct assault missions to the moon are deemed
necessary. Scheduled for earth-orbital flights, 1965 (APOLLO A);
circumlunar flights, 1966 (APOLLO B); and manned lunar landing 1966 or 1967 (APOLLO C). Project given $1 million funding
in FY 1961; $160 million, FY 1962. North American Aviation

(prime) awarded $400 million contract for 10 command modules and 10 service modules (Model A). Guidance navigation system to be developed by M.I.T.-Instrumentation Laboratory ($4 million contract for one year's work, to October 1962).

APT (Air Force)

Automatic Programed Tools approach. Developed for the Air Force Air Materiel Command. Since, has been installed or ordered by many missile firms. A technique for the machining of complex missile parts by which 80 to 95% in skilled man-hours may be saved. A high-speed digital computer calculates the motions made by a numerically controlled machine tool in cutting metal components. Developed by M.I.T.

ARADCOM

ARmy's Air Defense COMmand.

ARC

Designated RVX-2. A program to study advanced re-entry concepts in development of maneuverable nose cones. Unmanned. Basically an RVX with fins.

ARCADE (ARPA)

Continuous review of basic research areas and sparking of program actions. Program is associated with SUNRISE.

ARCAS (Navy)

High-altitude and meteorological sounding rocket. Length, 6 ft. Weight, 72 lb. Single stage. Capable of lofting 40-lb payload to a 70-mile altitude. Operational in 1960. Atlantic Research.

ARCAS ROBIN (Air Force)

Sounding rocket capable of lofting 8-lb payload to a 40-mile altitude. Atlantic Research.

ARCHER

"Aluminized" version of ARCON rocket. Capable of lofting 40-lb payload to over 80-mile altitude. Atlantic Research.

ARCMOLD

Plastic foams, both rigid and flexible, with densities from 1 to 40 lb/ft^3. Material has better insulating qualities than cork from -65° to 212°F. Impervious to water, humidity. Fungus- and bacteria-resistant. Atlantic Research.

ARCON (Navy)
High-altitude sound rocket. Solid propellant. Carried 40-lb payload to about 70 miles. BuOrd to use it for tests on vibration forces on solid-fuel rockets. Also on NASA's "approved" list. Operational in 1960. Atlantic Research.

ARCTURUS
Booster proposed by Martin in 1960 as successor to SATURN and replacement for NOVA, consisting of two NOVA engines and seven TITAN first-stage tank assemblies. To be capable of taking several men one-way to the moon, or one man round-trip without refueling.

ARDC
Air Research and Development Command. Now AFSC, Air Force Systems Command.

ARENTS (ARPA)
ARpa ENvironmental Test Satellite. Program to determine effects of altitudes on materials, components, and subsystems. Three shots planned for the 1962 time period. ATLAS-CENTAUR booster, 22,300-mile orbits. In R & D by G. D.-Astronautics.

ARGMA
Formerly ABMA. Army Rocket and Guided Missile Agency, Redstone Arsenal, Huntsville, Ala. Reorganized by AOMC, January 1962, with all missile programs placed in 13 weapon system project offices.

ARGO SERIES (NASA)
A series of hypersonic test vehicles which have been developed into high-altitude test and research rockets. Versions are A-1, C-1, D-4, D-6, D-8, and E-5. Aerolab Development.

ARGO A-1 (NASA)
Single-stage research rocket used to carry large payloads to high altitudes and to test components under high-velocity conditions. Length, 21.25 ft. Weight, loaded, 10,335 lb. Payload weight, 1000 lb. Altitude, 175 to 220 miles. Speed, Mach 8.75. Solid propellant. Aerolab Development.

ARGO C-1 (NASA)
Multistage test vehicle capable of lofting research payloads to 300-mile altitude. Length, 40.42 ft. Weight, loaded, 5800 lb. Solid propellant. Speed, Mach 9.75.

ARGO D-4 (NASA)

Also called JAVELIN. Four-stage sounding rocket capable of lofting 50-lb payload to 1000-mile altitude. Length, 47 ft. Weight at take-off, approximately 7000 lb. Speed, Mach 18.5. Rocket consists of HONEST JOHN booster, two NIKE boosters, and an X-245 (Allegheny). Engines: See X-216; X-244. Aerolab Development.

ARGO D-6 (NASA)

Four-stage research rocket designed to loft scientific payloads to 1500-mile altitude. Length, 55.75 ft. Weight, loaded, 8090 lb. Speed, Mach 19.5. Solid propellant. Aerolab Development.

ARGO D-8 (NASA)

Also called JOURNEYMAN. A four-stage rocket used to probe radiation belt. Capable of lofting 20-lb payload to 2800-mile altitude, or 50-lb payload to 2000-mile altitude. Length, 56.42 ft. Weight, loaded, 12,110 lb. Speed, Mach 24. Solid propellant. Rocket is composed of SERGEANT booster, two LANCES (Grand Central), and an X-248. Aerolab Development.

ARGO E-5 (NASA)

Also called JASON. Five-stage research rocket used in Project JASON to monitor the effects of nuclear explosions outside the earth's atmosphere; also explosions made by nuclear bombs in Project HARDTACK at Eniwetok and in Project ARGUS in the South Pacific in 1958. Length of rocket, 57.58 ft. Weight, loaded, 7250 lb. Payload weight, 60 lb. Altitude, 500 miles. Speed, Mach 12.5. Solid propellant. Rocket is composed of an HONEST JOHN booster, two NIKE boosters, a Thiokol RECRUIT engine, and a Thiokol T-55 engine. Aerolab Development.

ARGUS (ARPA)

High-altitude tests in the South Atlantic, August, September 1958, to monitor the effects of nuclear explosions. 4000-mile altitude. Three-stage X-17 rockets were launched from shipboard. Program was successful. Lockheed Missiles and Space.

ARIES (ONR/NASA)

Atmospheric Research Information Exchange Study. Study to determine the value of a technical information service on the upper atmosphere. Includes determination of the optimal technical coverage, staff requirements, service methods, and estimated annual operating cost of the service.

ARIES
> Authentic Reproduction of an Independent Earth Satellite. Full-scale model of an orbiting space laboratory presented to the American Museum of Natural History by Martin. Length, 41 ft. Diameter, 15 ft. Cost, $100,000. Model placed in the museum's Man-in-Space exhibition opened 12 October 1961.

ARKNIK (U.S.S.R.)
> Popular designation for U.S.S.R. SPUTNIK V, which contained the dogs Strelka and Belka as passengers. See SPUTNIK V.

ARM (Navy/Air Force)
> See SHRIKE.

ARMORNIGHT (Army)
> A human-factors study of personnel operating in thick-skinned weapons under low-visibility conditions.

ARPA
> Advanced Research Projects Agency.

ARPA 162-61 (ARPA)
> Formerly LOFTER. A satellite warning system with both defensive and offensive capabilities. Uses IR and ultraviolet sensing for detection of hostile ballistic missiles during their boost phase. Suspended and may be discarded.

ARPAT (ARPA)
> ARPA plus T for Terminal. Feasibility study to investigate possibilities for a new ballistic missile defense. Designed to meet re-entering ICBM warheads at the target end of their trajectories. ARPAT darts are to be spheroidal or ellipsoidal. May weigh less than 50 lb.

ARROW
> Rocket similar to LOKI. Used with a DEACON rocket to send chaff to over 300,000 ft. Sandia.

ARROWHEAD (ARPA)
> Term used for advance planning operation for ARPA projects, which culminate in recommendations to the Secretary of Defense.

ARSR-I-II
> Air Route Surveillance Radar developed for FAA by Raytheon.

ARTEMIS (Navy)
> An exploratory effort to track submarines at very long range by means of high-powered sonar transducers with high-gain receivers and data-processing equipment. This active sonar antisubmarine warfare system involves tests of a transducer suspended under a ship to produce high-intensity sound. In R&D. Part of TRIDENT. Columbia University, Hudson Laboratories.

ARTRON (WADD)
> Artificial neuron that will learn as a result of reward and punishment. This project to provide basic information needed for a thinking machine for use in automatic weather prediction, unmanned space exploration vehicles, and military game machines. Melpar.

AS-30 (France - Air Force/Navy)
> Air-to-surface missile similar to BULLPUP. Length, 12.42 ft. Weight, 1100 lb. Proportional control guidance. Solid propellant. Range, almost 9 miles, supersonic speed. HE or nuclear warhead. In evaluation trials by NATO. Developed by Nord Aviation.

ASA-31
> See JULIE.

ASCAMP
> Research vehicle capable of lofting 70-lb payload to a 150-mile altitude. Used with NIKE. Operational. Marquardt—Cooper.

ASCS (NASA)
> Automatic Stabilization and Control System. Used in MERCURY capsule.

ASD
> See AFASD.

ASFIR (Air Force)
> Active Swept Frequency Interferometer Radar. New type of radar to detect and track space vehicles. Technique, originated by Rome Air Development Center engineers, may provide resolution and position-measuring accuracy far higher than that from current radars 1963 is the target date for an experimental system.

ASIS (NASA)
> Abort Sensing and Implementation System. An electronic system to sense and predict failures which might endanger an astronaut's life. Weight of system, 36 lb. G. D.-Astronautics.

ASM
  Air-to-Surface Missile.

ASM-N-2
  See PETREL.

ASM-N-7
  See BULLPUP.

ASM-N-8
  See CORVUS.

ASP (Air Force)
  See AEROSPACE PLANE.

ASP (Navy)
  Atmospheric Sounding Project. Used in Operation REDWING.
  A single-stage rocket capable of lofting a 25-lb payload to a
  40-mile peak altitude. ASP I engine: See RM-1100; RM-1400.
  Rocket operational. ASP IV in development. Marquardt – Cooper.

ASPAN (NASA/Navy)
  Also called NIKE-ASP. Sodium-cloud research rocket capable
  of lofting a 70-lb payload to 145 miles. Used to determine the
  wind direction and velocity throughout the rocket's range and to
  measure rate at which matter disperses in upper atmosphere.
  Operational. Engine: See RM-1100; RM-1400. Marquardt –
  Cooper/Allegheny Ballistic Lab/Thiokol.

ASP/SCALE-SERGEANT
  Sounding vehicle capable of lofting 100-lb payload to a 20-mile
  altitude. Lower stage is an ASP; second stage is scaled-down
  SERGEANT.

ASROC (Navy)
  AntiSubmarine ROCket. Antisubmarine missile which is guided
  to target position, then eased into the water by parachute. It then
  seeks out the target and destroys it. Solid propellant. Uses active sonar to locate the target. Replaces RAT. Length, 15 ft.
  Launch weight, 1000 lb. Range, 8 nm. Speed, nearly Mach 1.
  Operational 1961. Minneapolis-Honeywell, prime.

ASSET (Air Force)
  Aerothermodynamic Structural System Evaluating Test. Program to launch six or seven hypersonic glider models to 400,000-
  mile altitudes at velocities of 19,000 to 20,000 ft/sec to study
  effects of heating on structures such as those used on DYNA-
  SOAR and follow-on boost-glide vehicles.

ASTER (Navy)

Antisubmarine torpedo which would be launched from the TERRIER missile. Range, 30 to 35 miles. ABL, prime; ABL/ Vitro, airframe; Ford-Aeronutronic, guidance; ABL, Rocketdyne, propulsion; Navy (in-house) warhead.

ASTOR (Navy)

AntiSubmarine Torpedo Ordinance Rocket. Underwater-to-underwater electric torpedo missile for ASW, to be launched from submarines. Length, nearly 20 ft. Weight, 2000 lb. Solid propellant. Range, 11 nm. Speed, nearly Mach 1. Nuclear or HE warhead. In R&D. Westinghouse, prime.

ASTRACON

Extremely sensitive electronic tube which can "see" individual particles of light. Used to photograph cosmic ray tracks. Westinghouse.

ASTRO (Navy)

Artificial Satellite Time and Radio Orbit. Navigational satellite.

ASTROBEE 500 (Air Force)

Two-stage sounding rocket capable of lofting a 40-lb payload to a 500-mile altitude. Series starts where AEROBEE ends. In R&D. Aerojet-General.

ASTROTUG

A general-purpose vehicle for performing coupling, maintenance, and repair in space. Vehicle to remain in orbit indefinitely, and be reprovisioned by fresh crews in re-entry vehicles. Proposed by Lockheed in 1960.

ASW

AntiSubmarine Warfare.

ATHODYD

AeroTHermODYnamic-Duct. A ramjet engine differing from the turbojet in that it has practically no moving parts – no compressor, shaft, or turbine. Engine needs only one-seventh of the fuel required by a rocket to develop the same thrust for the same period, since its oxygen is supplied by the atmosphere.

ATLANTIS (Navy)

A generic name for a group of ASW surveillance systems covering not only ARTEMIS, but also broad administrative functions associated with ASW projects. This program includes R&D for a 5000-mile range, solid-propellant ballistic missile if mounted

on surface vessels. If mounted on submarines and fired underwater, the range would be reduced. Expenditure through FY 1962,
about $52 million. See also TRIDENT, JULIE.

ATLAS (Air Force/NASA)
Military designations: SM-65; XSM-65; WS-107A-1; MX-1593.
Rocket-powered ballistic missile; A, B, C, D, E, F, G models.
Length, 80 ft. Weight, estimated at 200,000 lb. Speed, Mach 15.
Range, 5500 miles. Thrust, 370,000 lb. Liquid propellant. Guidance: first four squadrons, radio-inertial; last nine squadrons,
all inertial. "A" models tested boosters, guidance, and airframe. "B" models tested nosecone separation; sustainer
engine; guidance, including SCORE satellite; and operational
prototype. "C" models tested guidance, nosecone, and airframe
improvements. "D" models, operational type, 7500-mile range.
"E" models to be totally operational. "E" and "F" models
chosen to loft Air Force scientific payload-piggyback program;
22 flights scheduled for 1961 and 1962. "G" model is a proposed
booster with H-2 engines and a CENTAUR second stage. Has
up to 600,000-lb thrust. Engines: See LR-89 NA-5 MA-2 and
MA-3; LR-105 NA-5 MA-2 and MA-3. G. D.-Astronautics,
prime; G. E./Burroughs, Arma, guidance; Rocketdyne, propulsion; G. E./Avco, re-entry vehicle.

ATLAS-ABLE (NASA)
A three-stage space booster consisting of an ATLAS first stage
and ABLE upper stages. The ABLE stages were modified
VANGUARDS, consisting of an Aerojet-General 10-101 liquid-
propellant power-plant with a thrust of about 7500 lb. This combination is considered capable of sending 78 lb to Venus or 372 lb
into lunar orbit. Four firings made. One vehicle exploded during
static test and remaining three exploded in flight. See ABLE.
STL/G. D.-Astronautics.

ATLAS-ABLE IV (NASA)
Lunar probe with paddlewheels. Fired 26 November 1959, but
upper shroud came off 45 sec after lift-off. Payload and probably both upper stages were torn from the ATLAS booster.

ATLAS-ABLE V-A (NASA)
Unsuccessful lunar probe, fired 25 September 1960.

ATLAS-ABLE V-B (NASA)
Unsuccessful lunar probe, fired 15 December 1960.

ATLAS-AGENA A (Air Force)
Liquid-propellant space booster consisting of an ATLAS first

stage and an AGENA A second stage. Designed to launch MIDAS and SAMOS satellites. G. D.-Astronautics, first stage; Lockheed Missiles and Space, vehicle integration and second-stage structure.

ATLAS-AGENA B (Air Force/NASA)
Liquid-propellant space booster consisting of an ATLAS first stage and an AGENA B second stage. Capable of launching 5000 lb into earth orbit. Used to launch MIDAS II, April 1960. Scheduled to launch the 800-lb RANGER lunar probe. Operational. Lockheed Missiles and Space, prime; Bell, propulsion.

ATLAS-CENTAUR (NASA)
Space booster consisting of an ATLAS first stage and a liquid-hydrogen second stage. Length, 105 ft. Gross weight, 291,000 lb. Payload capability, 10,000 lb in 300-mile orbit or soft-land 730 lb on moon. First test flight scheduled for early 1962. Program delayed because of failure to attain simultaneous ignition of the two engines. Operational in 1962 or later. G.D.-Astronautics. May launch RANGER-type probes to Mars and Venus. In R&D.

ATLAS-HUSTLER (NASA)
Space booster capable of orbiting 3000-lb payload. Allied to MIDAS and TRIBE projects. Replaced by ATLAS-AGENA B. G. D. Astronautics/Rocketdyne/Bell.

ATOMIC CLOCK (NASA)
Project to test relativity theory of Einstein by comparing an atomic clock in a satellite with an atomic clock on earth. In R&D. Planned for 1961. Study contract by M.I.T.; contract for clock (with ammonia) to Hughes; contract for clock (with cesium) to NBS.

ATRAN (Air Force)
SSM surveillance. Version of MACE, designed for making maps. See also PINPOINT.

AUM
Air-to-Underwater Missile.

AUM-N
Air-to-Underwater Missile (Nuclear).

AURORA (Sweden)
Antisubmarine rocket launched from double-quadruple launcher at rate of eight rockets per minute. Maximum range, 850 m.

Speed, 70 to 90 m/sec. Detonation within 20 ft of submarine destroys it; within 50 ft, puts submarine out of action. Longer range rockets are LAURA, ERICKA, and FLORA. Deployed with Swedish Navy. Bofors.

AUTEC (Navy)
Atlantic Underwater Test and Evaluation Center, $372,000 six-month study to Lockheed-California, 1961.

AUTO-MET (Army)
Short-range, direct-support missile. Classified. Successor to Missile "A." Has guidance system to compensate automatically for weather and fly a true course. Included in MISSILE B project. ARGMA/G. E.

AZON (Air Force)
Military designation, VB-1. Unpowered guided bomb. Radio-controlled. Employed in China and Burma during World War II. See RAZON, TARZON.

AZUSA
Military designation, Mark II. An integration system which tracks all major missile shots electronically and predicts course and impact area. G. D.-Astronautics.

B

B-70 (Air Force)
Chemical-powered strategic bomber. Also called VALKYRIE; WS-110A. This supersonic intercontinental bomber system was initiated in 1955. NAA and Boeing were awarded study contracts. In December 1957, NAA was awarded the contract to develop WS-110A. In 1959 the contract was reduced to two prototypes, and in 1960 reinstated. In August 1961, a budget allocation of $448 million was approved for the B-70, as well as $525 million for manned bombers.

BABY ECHO (NASA)
Military designation, S-56. Air-density drag satellite, lofted by SCOUT early 1961. Langley Research Center, prime.

BABY S, T, R (Japan)
Series of rockets used by the University of Tokyo to obtain data on solid fuels, engine capabilities, and telemetry and recovery equipment. BABY S used primarily to train personnel in methods

of handling, launching, and tracking. Rocket constructed mainly of plastic, with aluminum tail fins. Initial firing over the ocean, August 1955. BABY T tested telemetering equipment. Six BABY T's fired in 1955. BABY R used in the investigation of ocean-recovery methods. Two rockets successfully recovered, 1955.

BADGE (Air Force)

Military designation, WS-212L. Large-scale system management, data-processing and display equipment. Designed to operate with detection and communications subsystems furnished by the government. G. E.

BALLISTA (Army)

The LOBBER missile which has been equipped with a warhead. Range, 10 to 15 miles. See also LOBBER.

BAMBI (Air Force)

Project proposed by RAND to ARPA. Calls for large number of satellites (800 to 3600) in low orbit providing global coverage. Satellites would carry ICBM interceptors capable of detecting and destroying ICBM's during their boost phase. Studies awarded to G. D.-Astronautics, STL, and Hughes, 1961. Contractors recommended construction of experimental hardware at a cost of about $40 million.

BASEBALL

A computer program in R&D by M.I.T. which can find in stored data on baseball games, the answers to questions fed into the computer in ordinary English.

BAT (Navy)

Radar-guided pilotless aircraft carried under the wings of big patrol bombers during the World War II period. Discontinued.

BAZOOKA (Army)

World War II rocket-propelled grenade which could be fired from an M-1 rifle as an antitank weapon. Developed in 1941 by a small rocket unit within the Army Ordnance Dept. Produced by G. E., and used in Tunisia, 1943. Number of variations of the BAZOOKA were developed as the war progressed. Some were built as one long tube (M1A1); others folded (M9A1).

BB-10 (France)

Air-to-surface missile. Solid propellant. Speed Mach 1 Radio-commanded. Carries TV camera so that pilot can guide missile to target. Operational. Deployed with French Air Force, Sud-Aviation.

BDM (Air Force)
Bomber Defense Missile. Military designation, WS-126A. Air-to-air or air-to-surface missile to defend bombers against ground or air attack. Adapted from the HAWK. Feasibility studies by G. E., Raytheon, and Hughes. Discontinued once, but believed again under consideration.

BEACON (FAA)
Aviation task force reporting to the President. Study of air traffic control and recommendation of a system for safe and efficient use of airspace. Richard R. Hough, Ohio Bell Tel. Co., Chairman. See also HORIZON.

BEACON (NASA/JPL)
Inflatable plastic satellite sphere launched 23 October 1958, JUPITER C booster. Mission: study of atmospheric density at various levels. Satellite failed to orbit because top stages separated prematurely while booster was still burning.

BEAR (U.S.S.R.)
Heavy bomber adapted to carry 50-ft-long winged missile with Mach 2 speed. Weight of BEAR, over 350,000 lb. Range, more than 9000 miles. Speed, approximately 400 mph. Turboprop-powered.

BEAUTY (U.S.S.R.)
Mach 2 bomber capable of attacking over intercontinental ranges in spite of existing air-defense systems. Possible 4000-mile range and high penetration ability. Ten of these bombers seen in the Tushino Show, 9 July 1961.

BELIER (France)
Single-stage, high-altitude sounding rocket. Can lift 40-lb payload to 50-mile altitude. Nonguided. Solid propellant. Speed Mach 4.8. Cost, approximately $6,000 each. Two rockets fired in the Sahara, May 1961. BELIER is one of a family of inexpensive sounding rockets being developed by Sud-Aviation. See AIGLE, CENTAURE, DRAGON, ERIDAN, PEGASE.

BELL HUSTLER
See HUSTLER.

BERENICE (France)
An improved ANTARES research rocket capable of Mach 12 re-entry. First launchings scheduled before end of 1961.

BETSY
See LULU.

BETTY (Navy)
Air-to-underwater or ship-to-underwater nuclear depth charge. Also called MARK 90. Offshoot of LULU. Operational.

BIG BOY (Air Force)
50,000-lb igniter developed for PROJECT 3059 and other million-lb-thrust class solid boosters. Aerojet-General.

BIG BROTHER
See DISCOVERER series.

BIG DISH (Navy)
Radio telescope. Diameter, 600 ft. Located at Naval Radio Research Center, Sugar Grove, W. Va.

BIG JOE (NASA)
An ATLAS-D booster used to loft the MERCURY capsule in unmanned shots, to determine structural integrity and heating, September 1959.

BIG SHOT
Project to observe separation, inflation, and possibly re-entry of the ECHO balloon by a TV camera mounted on the front of the THOR booster which launched the balloon. A post-burnout control system in the booster will keep the lens trained on the balloon.

BIG STICK (Air Force)
Nuclear-ramjet missile similar to BOLO. G.D.-Astronautics.

BIG WHEEL
Large land vehicle with wheels of 50-ft diameter, developed by Avro Aircraft to house workmen, equipment, workshops, and heavy machinery. Length of vehicle, 100 ft. Capable of carrying a 200-ton payload at 35 mph. Powered by sixteen 750-hp engines. Especially adaptable to developing remote areas of the world, including remote bases or launching sites. Avro Aircraft Co., Ltd., Canada.

BIMBO (Air Force)
Solid-propellant engine. Part of $4 million research program in large, solid-grain propulsion field. Thrust, 9 million lb. In R&D by Aerojet-General. Company-funded.

BIOS I (AEC)
Second launch in the nuclear emulsion recovery vehicle, NERV, scheduled for late 1961 by ARGO D-8 four-stage booster. Payload to contain biological specimens sensitive to various ranges of radiation, a cell-fertilization experiment, and one designed to study micrometeorite flux and penetrability. Aerolab Development.

BIRD DOG
See GENIE.

BIRDIE (Army)
Battery Integration and Radar DIsplay Equipment. A new transistorized version of the MISSILE MASTER, but in miniature. Installed at Turner AFB, Georgia. Used for coordination and control of NIKE AJAX and NIKE HERCULES batteries. Martin Orlando. See MISSILE MASTER.

BISON (U.S.S.R.)
Four-jet heavy bomber first exhibited May Day, 1954. Weight, loaded, about 350,000 lb. Range, about 6000 miles. Speed, approximately 550 mph. Length, 160 ft. Wing span, about 170 ft.

BLACK ARROW (Air Force/NASA)
See X-15.

BLACK BRANT III, IV, V (Canada)
Solid-fueled, single-stage rocket family designed to lift 150-lb payload to 150- to 200-mile altitude. BLACK BRANT III: One-stage rocket. Maximum diameter, 10 in. BLACK BRANT IV: Two-stage rocket. Maximum diameter, first stage, 17 in; second stage, 10 in. BLACK BRANT V: One-stage rocket. Maximum diameter, 17 in. All three versions use the solid propellant developed by CARDE. Scheduled launch of prototypes from Fort Churchill Rocket Range, 1962. Bristol Aero-Industries, Ltd., Canada.

BLACK KNIGHT (Great Britain, Royal Aircraft Establishment)
High-altitude research rocket, one-stage and two-stage versions. Development initiated in 1955. Now operational. Used for testing nosecones and components of the BLUE STREAK booster. Single-stage version: Length, 35 ft. Diameter, 3 ft. Maximum velocity at burnout, about 8000 mph. Capable of lofting payload of about 20 lb. Two-stage version: First-stage, Bristol Siddeley Gamma 2 rocket engine. Thrust, about 16,400 lb at sea level. Weight, about 700 lb. Liquid propellant. Second-stage, solid or

liquid. Altitude, 300 to 500 miles. Proposed for use for space probes, employing BLUE STREAK as the first stage and BLACK KNIGHT as the second stage. Saunders-Roe/Bristol Siddeley.

BLINDER (U.S.S.R.)
Interceptor developed to meet threat of SAC's B-52—HOUND DOG missile combination. Capable of intercepting the B-52 more than 1000 miles from Russian targets, thus preventing bomber from getting close enough to launch the HOUND DOG. Also capable of subsonic speeds in the interception zone for long periods before or between combats.

BLOODHOUND I (Great Britain)
Air-breathing, surface-to-air missile. Length, 28.3 ft. Launch weight, 4500 lb. Solid propellant. Range, approximately 35 miles. Speed, Mach 2.3. Radar-homing guidance. To be located with combat units in Britain and Australia. Also sold to Sweden. BLOODHOUND II under development. Will have antimissile capability. Bristol Aircraft.

BLUE GRASS (Air Force)
Classified communications project. $5.8 million, budget for FY 1961.

BLUE SCOUT I, II (Air Force)
Solid-propellant research rocket almost identical to SCOUT except for (1) destruct package and controls, (2) a 30-in. diameter payload carrier instead of one 20-in., and (3) fourth stage sometimes eliminated (BLUE SCOUT I) or replaced by Aerojet or NOTS rocket (BLUE SCOUT II). Rocket originated in HETS program. Used for high-altitude sounding, aerodynamic, subsystem warhead testing, and satellite launching. Especially planned to test DYNA-SOAR. Capable of carrying DYNA-SOAR model of 500 lb to 300,000 ft at up to 20,000 ft/sec. Also scheduled for use in TRANSIT and SAINT programs. About 35 BLUE SCOUT firings planned annually. Cost per copy, about $750,000. Ford-Aeronutronic, prime; Ling-Temco Vought, vehicle. Minneapolis-Honeywell, guidance; Aerojet/Hercules/Thiokol, propulsion.

BLUE SCOUT JR (Air Force)
Military designation, TS-609A. Four-stage, high-altitude probing vehicle. First stage same as second stage of SCOUT. Height, 40 ft. Weight, 14,000 lb. Maximum altitude, 20,000 miles. Maximum velocity, 20,000 ft/sec. Solid propellant. Thrust, 50,000 lb. Smallest of the three configurations making up the HETS 609A

vehicles. Eleven shots scheduled for 1961. Has carried a pay-load of 32.8 lb to 16,600-mile altitude, with a range of 7000 miles.

BLUE STEEL (Great Britain)
Air-to-surface tactical missile. Length, 36 ft. Weight, about 15,000 lb. Liquid propellant. Range, approximately 500 miles. Speed, Mach 12. Nuclear or HE warhead. Air-launched from British V-bombers. Can feint and maneuver evasively during flight to an assigned target area. Deployed with Royal Air Force. A. V. Roe.

BLUE STREAK (Great Britain)
Surface-to-surface ballistic missile in the IRBM class, initiated in 1954. Length, 70 ft. Diameter, 10 ft. Two RZ 2 Mk 2 liquid-propellant (gasoline and liquid oxygen) engines with 135,000-lb thrust each. Test firings have been made. Engine based on the MA-2 (ATLAS), MB-3 (THOR), and S-3 (JUPITER) engines. Cost, about $16 to $23 million per year, until 1959 when costs rose to about $33 million. Program canceled in 1960, but booster development was continued. Proposed to the European Space organization as a space booster for cooperative efforts. De-Havilland.

BLUE WATER (Great Britain)
Surface-to-surface missile which can be hauled by its own transporter-erector on a 3-ton truck. Has raked tips on for-ward cruciform wings. Length, 25 ft. Solid propellant. Inertial guidance. Nuclear or HE warhead. Range, about 80 to 100 miles. To be deployed with British combat troops. Also sold to West German army. Operational in 1963. English Electric.

BMD
See AFBMD.

BMEWS
Ballistic Missile Early Warning System. Network of long-range radar stations to cover the Soviet-Sino bloc from the Pacific to Poland. To be fully operational by 1962. Cost, $700 million.

BO-4
See ROBOT 304.

BOA (Air Force)
Sounding rocket capable of carrying 500-lb payload to a 75-mile altitude. Length, 39 ft. Weight, 6666 lb. HONEST JOHN, first stage; NIKE, second and third stages. Marquardt–Cooper.

BOAR (Navy)
Air-to-surface unguided rocket. Nuclear warhead capability. Designed for supersonic planes making low-altitude nuclear strikes to escape radar detection. Royal Industries.

BOBBIN (Great Britain)
Test vehicle for the BLOODHOUND surface-to-air missile. Solid-propellant booster. Used also to check out vehicle recovery at supersonic speeds. Various recoverable versions produced.

BOLA
Concept for manned space vehicle in which a long cable separates the two parts of the vehicle. Could be used to separate a reactor from a manned vehicle.

BOLD ORION (Air Force)
Military designation, WS-169A. Early name for SKYBOLT program.

BOLO (Air Force)
Nuclear-ramjet missile. Proposal by North American.

BOLT
See M-55.

BOMARC A,B (Air Force/RCAF)
Air-breathing, surface-to-air missile. Length, 47 ft. Weight, 15,000 lb. Liquid propellant. Command guidance, active radar homing. Mach 3 at about 60,000 ft. Range, 250 to 400 miles. Thrust, 270,000 lb. B model in production. In March 1961, BOMARC B intercepted simulated supersonic target at a 100,000-ft altitude and 446-mile range. For BOMARC A: Boeing, prime; IBM/Westinghouse, guidance; Aerojet/Marquardt, propulsion. For BOMARC B: Boeing, prime; Kearfott/Westinghouse/IBM, guidance; Thiokol/Marquardt, propulsion.

BOMBARDMENT (Air Force)
Air-to-air missile for use by long-range bombers, such as the B-52. Being developed by North American.

BOMBER-58 POD (Air Force)
Missile to be launched from the B-58 Hustler bomber. Status, uncertain. Bell Aircraft.

BOMI (Air Force)
Also BRASS BELL. Original Bell Aircraft proposal for a

boost-glide vehicle, as DYNA-SOAR. (Boeing and Martin awarded the DYNA-SOAR contract.)

BOSS (Air Force)
Bioastronautical Orbital Space System. Plan to use a Lockheed-built Advanced Biomedical Capsule (ABC) to orbit large chimps for 14 days at a time. Mission: to provide data on long-term weightlessness and radiation, by means of internal biomedical transducers. Program not yet carried out.

BOSS-WEDGE (Air Force)
Bomb Orbital Strategic System. Three-year study program by Boeing to assess value of using manned and unmanned space vehicles in bombing, reconnaissance, and other military missions. Outgrowth of company-funded project WEDGE.

BOUNDER (U.S.S.R.)
Soviet delta-wing bomber which appeared at air show July 1961. Superior to B-58. Believed capable of about Mach 2 with use of afterburner.

BRASS BELL
Name given boost-glide vehicle in early Bell Aircraft proposal. See BOMI, DYNA-SOAR.

BRAVO
See ALPHA ABC.

BSD
See AFBSD.

BTZ-411.01 (France)
A surface-to-surface, two-stage missile, similar to a bazooka. Coleopter configuration. BTZ.

BUG (Air Force)
Papier mâché missile, using a preset guidance system. Developed during World War I by a group of American scientists and engineers (Orville Wright, C. F. Kettering, and Elmer Sperry). Flight-tested successfully in fall of 1917. Project continued until 1925. Improved 200-mile-range version was developen and flight-tested in 1941, but was dropped.

BULLDOG (Navy)
Air-to-surface missile. Mission: to destroy small targets. An advanced version of BULLPUP believed to be in late phases of R&D. Classified. Martin.

BULL GOOSE (Air Force)
See GOOSE.

BULLPUP (Navy/Air Force-TAC)
Military designations: GAM-83; TGAM-83; ASM-N-7A,B; WS-321A. Air-to-surface missile capable of being launched from present and future light-attack aircraft, or from carrier or shore-based planes. Length, 11 ft. Weight, 540 lb. Radio-commanded. Range, 15,000 ft. Speed, Mach 2. BULLPUP A in production. BULLPUP B in late development. Navy and Air Force A models are identical. Navy model B has larger HE warhead. Air Force model B has nuclear warhead. Martin, prime, Maxon, second-source prime.

BUMBLEBEE (Navy)
World War II program at Johns Hopkins University to find defense against Japanese kamikaze attacks on the U.S. fleet. Program begun 1944. Laboratory developed the TALOS as the fleet's defense. Major armament of cruiser Galveston, other cruisers, since 1958. See TALOS.

BUMPER (Army)
Also BUMPER WAC. Program in which eight combination rockets (ex-German V2 and WAC CORPORAL) were fired between 1948 and 1950, contributing research data to VANGUARD and the ICBM. Douglas, G. E., JPL.

C

C-1
Military designation for long-range version of Navy POLARIS.

C-1, C-2, C-3 (NASA)
Three, four, and five-stage models of SATURN booster. Chance-Vought, Douglas, G. D.—Astronautics.

C-2 (U.S.S.R. – Air Force)
Surface-to-air missile. Length, 25.8 ft. Weight, 8400 lb. HE warhead. Beam-rider guidance. Range, 25 miles at Mach 2.0. Ceiling, 50,000 ft. Missile was developed from World War II German missile, WASSERFALL.

C-7 (Italy – Air Force)
Air-to-air missile with six-mile range. Speed, Mach 1.9. IR guidance. Solid booster. Deployed aboard Italian Air Force interceptors. Later models to have alternative radar guidance

and Italian booster (Bombrini Parodi-Delfino). Nearly operational. Società Italiana Sviluppo Propulsione a Reazione.

CAESAR (Navy)

Submarine detection system. Encapsulated sonar devices dropped in ocean. Part of TRIDENT. Operational 1959.

CAISSEUR (France – Army)

Air-breathing surface-to-surface tactical missile, with 16- to 60-nm range. Speed Mach 0.8. Solid booster, ramjet engine. Radar-command guidance. Launched from mobile carrier. Operational. Sud-Aviation.

CAJUN (Army/Navy)

Research vehicle composed of NIKE booster and DEACON second stage. Solid propulsion. Replacement for the ROCKOON. Model III was designed for use in the POGO-HI target missile program. Operational; no production. Thiokol.

CALEB (Navy)

Also called NOTSNIK in former project. Reactivated in 1960 at Pt. Mugu, Calif. Air-launched orbital satellite. Length, $16\frac{1}{2}$ ft. Diameter, 2 ft. Weight, 3000 lb. Maximum speed, 18,000 mph. Solid propellant. Fired from zoom-gliding F4H Phantom at about 80,000 ft in program similar to Air Force FARSIDE, using aircraft instead of balloons. Tracking done at Arguello. Technique developed at China Lake under YO-YO. No launch attempted under CALEB. See also NOTSNIK, NOTS.

CAMAL (Air Force)

Continuously Airborne Missile launcher And Low-level penetration bomber. Concept calling for 40 out of 85 nuclear aircraft continuously in the air, each carrying five ALBM (SKYBOLT). Program canceled March 1961.

CAMP (U. S. Patent Office)

A punched-card system using phosphorus for mechanized search.

CAMP

Cost of Alternative Military Programs. An operations research and long-range planning system to relate requirements to time, cost, and design. G. E.—Tempo.

CANNONBALL (Navy, BuOrd)

Military designation, (D-40-2)SSM. Antitank or antipersonnel, solid-propellant missile.

CANTALOUPE (Air Force)
Missile defense system study. Simulation of nuclear explosions in space. Air Force version of BAMBI. In study phase. Republic Aviation.

CARDE
Canadian Armament Research and Development Establishment. Element of the Canadian Defense Board, charged with providing advice and research on weapon systems.

CARDINAL
See KATYBIRD KDB-1.

CARIBE (believed Navy)
Classified antisubmarine weapon. Reportedly underwater-to-underwater and surface-to-underwater missile. Early research.

CASDN (France)
Comité d'Action Scientifique de la Défense Nationale.

CASTOR (NASA)
Second stage of SCOUT booster. Modification of the SERGEANT. Length, 20 ft. Weight, 8800 lb. Thrust, more than 50,000 lb. Used clustered in LITTLE JOE program of MERCURY. See SCOUT. Thiokol.

CAT
Experimental missile. In R&D.

CATE
Current AFSC Technical Efforts. Project to prepare index of nearly 50,000 scientists and engineers. Begun 1960. Intended as a source of technical information for scientists; also as aid in making management decisions and speeding up identification of scientists and engineers working in determined areas of interest to the Air Force.

CEBMCO (Army)
Corps of Engineers Ballistic Missile COmmand in Inglewood, Calif. Responsible to the Air Force for activating ICBM missile sites.

CELESCOPE (NASA)
Experiment planned for the Orbiting Astronomical Observatory (OAO). Will use the UVICON which will respond to the short-wave ultraviolet rays emitted by celestial bodies but never

reach the earth. The UVICON to be launched in an AEROBEE-HI rocket to a 200-mile altitude to sample the strength and nature of ultraviolet radiations and get preliminary data on the performance of CELESCOPE. See also UVICON, OAO. Ling-Temco Vought.

CEMS

Central Electronic Management System. Function: to integrate all or many of the functions now performed on subsonic aircraft by variety of subsystems, and to improve safety and economy. Also to increase available airspace for improved traffic control. Hughes.

CENTAUR (NASA)

Two-stage space booster consisting of an ATLAS first stage and a liquid-hydrogen second stage. To be capable of placing about 8000-lb payload into 300-mile orbit or of soft-landing over 700 lb on the moon. Expected to be the first U.S. booster to exceed the weight-lifting capability of U.S.S.R. boosters which lofted, for example, the 1961 Venus probe. CENTAUR is scheduled to be used for such programs as SURVEYOR, MARINER, and ADVENT. Convair, prime; Pratt & Whitney, propulsion; Minneapolis-Honeywell, guidance.

CENTAURE (France)

High-altitude sounding rocket composed of BELIER rocket with first-stage booster. Will loft 40-lb payload to 110-mile altitude. Speed, Mach 5.5. Cost, approximately $10,000 each. Two CENTAURES were fired in the Sahara Desert, May 1961. One of a family of inexpensive rockets being developed by Sud Aviation. See AIGLE, BELIER, DRAGON, ERIDAN, PEGASE.

CERMA

Centre d'Etudes et de Recherches de Médecine Aéronautique. French Center of Aeromedical Studies and Research.

CERN

Centre Européen de Recherches Nucléaires. European Organization for Nuclear Research. Organized in the early 1950's. Headquarters in Switzerland.

CETA (France)

Centre d'Etudes pour la Traduction Automatique. A CNRS foundation partly supported by CASDN.

CETIS
CEntre de Traitement de l'Information Scientifique. Scientific Information Processing Center, Brussels, Belgium. Includes: GRISA (Scientific Data Processing Research Team), the original founders of CETIS and now the research part of CETIS; and DOCA (Automatic Documentation).

CH-17
See KOMET 1 (U.S.S.R.).

CH-18
See KOMET 2 (U.S.S.R.).

CHARIOT (AEC, CRREL)
Cratering experiment at Cape Thompson, Alaska, north of the Bering Strait, that may demonstrate feasibility of using nuclear explosives for large-scale excavation, e.g., artificial harbor on the arctic Alaskan coast. See CRREL.

CHARIOT
Nickname given high-energy, liquid-propellant upper stage to be developed for use on the TITAN II as a backup vehicle for ADVENT launchings.

CHARM (Canada)
Three-stage sounding rocket. Length, 71 in. Diameter, 3 to 6 in. Capability, 100,000-ft altitude.

CHEAPIE (Air Force)
Proposed anti-ICBM. Range unknown. Little support. Feasibility studies.

CHEROKEE (Air Force)
Solid-propellant missile. Thrust, 50,000 lb. Tests ejection-seat development. Operational. Cook Electric.

CHOPPER JOHN (Army)
Helicopter-aided version of HONEST JOHN missile. Part of 1958 testing project at White Sands Proving Ground. Missile moved into position by helicopter, but ground-launched.

CLAM (Air Force)
Ramjet-powered supersonic missile. Chemically fueled. Low altitude. Project to achieve SLAM-type capability before SLAM can be developed. Same design range as SLAM, but shorter range capability. In research.

CLAYMORE (Army)
Classified project, reportedly an antipersonnel missile. In R&D.

CLINKER (Navy)
Long-wavelength IR detectors capable of sensing heat differences in submarine water trails. Test of the theory that a submarine moving deep in the ocean causes a change in the temperature at the surface. Goodyear Aircraft.

CNRS (France)
Comité National pour les Recherches Scientifiques.

COBOL
COmmon Business Oriented Language. Interchange of computer programs utilizing plain English (rather than machine code) between data-processing systems of different manufacturers. RCA/Remington Rand.

COBRA (Marines)
Antitank and antivehicle attack system. Initially developed in Switzerland. Made in West Germany. Extensive use of plastics. Solid propellant. Wire guidance. Speed, 191 mph. 5.5-lb HE warhead. Operational with West Germans. Recommended for purchase by U.S. Marine Corps. Can be launched and guided by one man. Boelkow Entwicklungen, Munich, Germany, and Daystrom, Inc., U. S. licensee.

COBRA (Navy)
Air-launched antiship radar missile. Short-range version of CORVUS. Early R&D. No contract announced.

COC
COmmand and Control system for the North American Air Defense Command's underground combat center being built in Cheyenne Mountain near Colorado Springs, Colo.

COCA
See ALPHA ABC.

CODIT
COmputer DIrect to Telegraph. Transmits predicted trajectory data at Cape Canaveral. Has capability for world-wide transmission.

COLIDAR

COherent LIght raDAR. A lightweight, low-power, small-sized radar for use in identifying objects in space. Can isolate a single target within a group of closely spaced targets. Hughes.

COMANCHE

Solid-propellant research and test vehicle in R&D.

COMET 1, 2, 3 (U.S.S.R.)

See KOMET.

COMIT (National Science Foundation)

A programing system for computers. M.I.T.

COMLOGNET

Transceiver and teletype communications system for ballistic missile logistics. To replace LOGBALNET system, July 1961. COMLOGNET can handle all logistic communications traffic without building a message backlog.

COMSAT

Abbreviation for communications satellite.

CONALOG

CONtact anALOG. A display on a screen of a submarine navigation corridor which appears as a tunnel. Simulation of the proper course with the bottom and the top of the ocean. Tested on submarine Shark.

CONARC (Army)

U.S. CONtinental ARmy Command. Responsible for the ground defense of the U.S., Army training, reserve forces, ROTC, logistics, combat developments, and the Strategic Army Command.

CONTIN (Italy)

Liquid-propellant research rocket. Under development.

CONTRAVES (Italy)

Liquid-propellant surface-to-air missile. Under development.

COPERS

COmmission Préparatoire Européenne de Recherches Spatiales. Group of 11 European nations for space research. Secretary-General, Prof. Pierre Auger (France). Commission's task is to lay the groundwork for establishment of the European Space Research Organization (ESRO).

CORPORAL (Army)
Surface-to-surface field artillery missile. Length, 45 ft. Weight, 12,000 lb. Guidance, radio-command; inertial. Speed, Mach 3. Range, 75 miles. Liquid propellant. Thrust, 20,000 lb. Nuclear or HE warhead. Deployed with U.S. and NATO troops in Europe. Initially operational, late 1953. To be replaced by the SERGEANT. Firestone, prime; Gilfillan, guidance; Ryan, propulsion.

CORVUS (Navy)
Rocket-boosted air-to-surface missile. Military designation, XASM-N-8. Range, in excess of 100 miles. Supersonic speed. Canceled. Engine: See TD-165. Ling-Temco Vought.

COSMOS
Six million-lb thrust, liquid-propellant rocket booster. Plug nozzle. Length, 149.5 ft. Weight, 350,000 lb. Proposed by Aerojet-General.

COSPAR
COmmittee on SPace and Aeronautical Research. An international space organization which came into being in 1959 and held its first international Space Science Symposium in 1960, now an annual event.

COURIER (ARPA/Army)
Communications satellite. Weight, 475 lb. Diameter, 51 in. Mission: to test feasibility of global military communications network using "delayed repeater" satellites which receive and store information until commanded to transmit. COURIER 1A failed 18 August 1960 when its THOR ABLESTAR booster blew up. COURIER 1B was launched successfully into an orbit approximately 650 nm, but performed less than three weeks. Program is successor to SCORE. Part of NOTUS project. Estimated to be operational in 1962 or 1963.

COZI (ARDC)
COmmunications Zone Indicator.

CPFF
Cost Plus Fixed Fee. For government contracts.

CPIF
Cost Plus Incentive Fee. For government contracts.

CREE (Air Force)
Supersonic, three-pronged rocket test missile. Tests three parachutes at a time. Has capability of lofting 800 lb to a 200-mile altitude at Mach 2. Operational. Cook Electric.

CRICKET
Cold-propellant meteorological rocket, capable of being launched by only one man. Recoverable by parachute. Has lowest cost per flight ($6). Capable of reaching 3000-ft altitude. Texaco Experiment.

CRONUS
Part of a Grand Central Rocket proposal for a super solid booster capable of placing 26,000 lb in a 300-mile orbit. To have total weight of 956,000 lb.

CROSSBOW (Air Force)
Winged air-to-surface missile. Range, 200 miles. Speed, 500 knots. Adaptation of the drone aircraft to combat use. Canceled. Radioplane.

CROW
Warhead for the ZUNI missile developed and evaluated by NOTS in conjunction with the New Mexico School of Mines. Rheem-Aerojet.

CROW (Navy)
Air-to-air missile. Solid propellant. In R&D.

CRREL (Army)
Cold Regions Research and Engineering Laboratory. Formerly SIPRE (Snow, Ice, and Permafrost Research Establishment). See CHARIOT.

CSAR (Air Force)
Formerly FLAG. Military designation, SR-178. Communications Satellite Advanced Research. A synchronous 24-hr communications satellite. Long-range strategic communications. Solar cell power. Several duplex voice channels. Weight, 2 tons. Follow-on for ADVENT. To be operational by 1965. ATLAS-CENTAUR probable launch vehicle.

CT. 20
Subsonic target missile, turbojet engine, command guidance. Parachute recoverable. Speed, 560 mph at 33,000-ft altitude. Endurance, 45 min. Bell Aircraft (mfg rights from Nord Aviation).

CT. 41
Target drone more advanced than the CT. 20. Speeds up to 1650 mph. Designed for weapon system training and evaluation. In quantity production in 1961. Bell Aircraft (mfg rights from Nord Aviation).

CTV
Abbreviation for Control Test Vehicle.

CUTWATER (Navy)
Project to study nonacoustic means for detecting submarines, i.e., IR, electromagnetic, and other approaches.

CYBERTRON
Designation: K100; K200. A learning machine which duplicates the human learning process, experience, trial and error, correlation of new facts with past experience, and the use of a human teacher. Similar to the Cornell Aeronautical Laboratory PERCEPTRON. Developed by Raytheon.

D

(D-40-2)SSM
See CANNONBALL.

D-58
See SKYROCKET.
S
D-558-II
See SKYROCKET.

DAC (Defense Atomic Support Agency)
Damage Assessment Center. Contracts for interservice DAC design and assessment of nuclear effects awarded to System Development Corp. for computer model development and programing; to Control Data Corp. for computer system; and to Ramo Wooldridge for display system.

DACOM
DAtascope Computer Output Microfilmer. Device permitting magnetic tapes from electronic computers to be read and transferred immediately in high-quality characters to microfilm. Kodak.

DAEDALUS
See MAN-HIGH.

DAMP (ARPA/Army)

Down-range Antimissile Measurement Program. Part of Project DEFENDER. Optical instruments used to investigate reentering missiles and also to track satellites for NASA, especially ECHO. RCA/Barnes Engineering.

DAN (Air Force)

Discontinued project. Experimental liquid-fueled missile. Contraction of Deacon And Nike. Replaced by NIKE-CAJUN. Developed for ARDC Cambridge Research Center. University of Michigan/NACA.

DAN (Navy)

Two-stage sounding rocket capable of lofting a 50-lb payload to a 75-mile altitude. Marquardt-Cooper.

DARB

Distressed Airman Recovery Beacon. Small lightweight beacon to be worn by pilot. Generates distress signals on which rescue planes can home directly. Hallamore Electronics.

DARK FENCE (Navy)

A designation for SPAR, a space surveillance system which stretches across Southern United States.

DART (Army)

Surface-to-surface antitank missile. Wire command. Solid propellant. Speed, 900 ft/sec. Range, 2 miles. Length, 5 ft. Weight, 100 lb. For use by infantry or armored units. Program initiated March 1952; canceled September 1958. Nord Aviation's SS-10 and SS-11 adopted instead. See also ENTAC. Aerophysics/Curtiss-Wright.

DART (Navy)

Research rocket used for gathering weather information. Unguided. Fired from 5-in. guns on shipboard. Used in HASP program. Under development by the Naval Ordnance Laboratory.

DASH (Navy)

Drone AntiSub Helicopter. A remotely controlled drone helicopter. Unmanned. To be flown from destroyers; drops payload on submarine target, and returns to parent ship. See GYRODYNE. Gyrodyne.

DAT (Navy)
Underwater-to-underwater drone. In R&D.

DATA-STOR
Ground portion of guidance computers, tape devices. Used with missile programs SERGEANT, POLARIS, and ATLAS. In production. Cook Electric.

DAVY CROCKETT (Army)
Portable, tactical nuclear rocket. Solid propellant. Designed for firing from bazooka tubes of different sizes for different ranges. Relatively short range. In R&D, to be operational in 1962. Manufactured at Army's Rock Island, Ill. Arsenal. Reportedly, Marine Corps has abandoned procurement plans for DAVY CROCKETT.

DEACON (Army)
Research rocket used for model testing by companies. Also used by NACA for research tests; and used in Operation POGO and Project ROCKOON. When launched from balloon, called ROCKOON; when launched from aircraft, called ROCKAIRE; when used with NIKE, called DAN. Over-all weight, 90 to 150 lb. Speed, 2000 ft/sec. Engine: See X-220. Hercules Powder.

DEACON-ARROW
Small, low-cost sounding rocket, capable of carrying a 12- to 20-lb payload to 100-mile altitude. Two-stage vehicle. Length, 17 ft. Solid propellant. Composed of Alleghany Ballistics Laboratory's DEACON in the first stage and Grand Central Rocket's ARROW in the second stage. Developed by Sandia along with VIPER-ARROW.

DECCA NAVIGATOR
A British navigation system used for aircraft and ships. Decca Navigator Co., Ltd., London.

DECREE (ARPA)
Equatorial 24-hr satellite. R&D under ARPA; booster launch under Air Force; operational program directed by Army Signal Corps. In R&D. G. E.

DEFA (France)
Direction des Etudes et Fabrications d'Armement. Government agency in charge of research and development of armament. See ENTAC.

DEFENDER (ARPA/Air Force/Army/Navy)
An interservice project under the over-all cognizance of ARPA. Aim of program is to increase scientific and technical knowledge of the problems of defense against space objects (including spacecraft and ballistic missiles), and thus to devise means of defense against attacks from space. DEFENDER covers general physical research and technological research, including instrumentation, ballistic missile phenomena, systems studies, characteristics of the upper atmosphere, radar development, and reentry body identification. In R&D. Includes such projects as ESAR, GLIPAR, LONGSIGHT, PINCUSHION, PRESS, SPAD, and TRADEX.

DEIMOS
DEvelopment and Investigation of a Military Orbital System. A Martin proposal to the Air Force for a manned space logistics system as an intermediate step to such systems as MTSS, SLOMAR, and SMART. Concept uses railroad standardization, with standard "freight cars," "passenger cars," and couplings. These could be assembled into a train after separate boosting into space. Booster would be a modified TITAN II.

DELAWARE
Solid-propellant research and test vehicle in R&D.

DELTA (NASA)
Three-stage vehicle capable of launching a 480-lb payload into a 300-mile earth orbit, or a 65-lb payload on a space-probe mission. Liquid-solid propellant. Successor to THOR-ABLE, DELTA is an interim booster for NASA shots. Goddard, prime; Douglas, booster; Bell, guidance; Rocketdyne/Aerojet/ABL, propulsion.

DELTA DART (Air Force)
Convair interceptor developed for the Air Force.

DERVISH (Army)
Surface-to-air solid-propellant missile. First developed to replace the air-to-air MIGHTY MOUSE. Reactivated in 1958. In R&D. Thiokol.

DEW LINE (Air Force)
Defense, August 1961, as a joint military intelligence organiza-
the width of the North American continent.

DIA (DOD)
Defense Intelligence Agency. Established by the Department of Defense, August 1961, as a joint military intelligence organization. Purpose: greater unification of military intelligence and elimination of duplication.

DIAMANT
See TOPAZE.

DIAMONDBACK (Navy)
Air-to-air missile. An advanced version of SIDEWINDER, with greater range and accuracy. NOTS development. Now an inactive program. Philco/Naval Ordnance.

DIANA-A
A liquid-hydrogen, liquid-oxygen second stage for a space booster. Proposed by Douglas to the Air Force and NASA. First stage, a THOR with three Thiokol 50,000-lb-thrust solid rockets clustered at its base.

DING-DONG
See GENIE.

DISCOVERER (Air Force/NASA)
Satellite research program aimed at stabilization of a satellite in orbit and recovery of the payload capsule by air snatch or from the ocean after flight. Series boosted into a polar orbit from Vandenberg Field, Calif. (PMR) by THOR vehicle. Program phasing out. For details, see specific DISCOVERER shots. Lockheed, prime; G. E., re-entry vehicle.

DISCOVERER I
Satellite launched 28 February 1959. First U.S. satellite to be fired into polar orbit. Length, 18.8 ft. Diameter, 60 in. Payload gross weight, 1450 lb; net weight, 1300 lb. Elliptical orbit. Inclination to equator, 89.96°. Perigee, 99 miles; apogee, 605 miles. Objectives were to check out propulsion, guidance, staging, and communications. Satellite tumbled badly. Remained in orbit to 5 March 1959 when it re-entered the atmosphere and burned.

DISCOVERER II
Satellite orbited 13 April 1959. Length, 19.2 ft. Diameter, 60 in. Payload gross weight, 1610 lb; net weight, 1440 lb. Circular orbit. Inclination to equator, 90.2°. Perigee, 152 miles; apogee, 220 miles. Objectives of flight were to recover capsule, maintain temperature and oxygen in sufficient amounts to sustain

life, and measure radiation by emulsion packs. Experiments were successful, including capsule separation from vehicle in orbit. However, ejected capsule was not recovered. In orbit to 26 April 1959, then re-entered over Spitzbergen instead of Hawaii and was lost.

DISCOVERER III

Satellite launched 3 June 1959. Weight, 1600 lb. Objectives were measurement of cosmic radiation, biomedical environmental research, and capsule recovery techniques by C-119 aircraft, patrolling recovery area. Tracking stations received no telemetry from satellite. Doubtful it achieved orbit since angle was too steep.

DISCOVERER IV

Satellite launched 25 June 1959. Gross weight, 1700 lb. Failed to achieve orbit because of insufficient velocity.

DISCOVERER V

Satellite launched 13 August 1959. Length and diameter same as DISCOVERER II. Gross weight, 1700 lb; net weight, 1440 lb. Near-circular orbit. Inclination to equator, 172°. Perigee, 136 miles; apogee, 450 miles. Satellite orbited until 16 September 1959, but retrorocket malfunction caused by re-entry capsule to stay in higher orbit.

DISCOVERER VI

Satellite launched 19 August 1959. Length and diameter same as DISCOVERER II. Gross weight, 1700 lb. Near-circular orbit. Inclination to equator, 172°. Perigee, 139 miles; apogee, 537 miles. In orbit until 20 October 1959. Capsule separated but was not recovered.

DISCOVERER VII

Satellite launched 7 November 1959. Gross weight, 1700 lb. Elliptical orbit. Inclination to equator, 81.66°. Perigee, 98.5 miles; apogee, 510 miles. Satellite failed to stabilize in orbit because of a malfunction of a 400-cycle inverter. Capsule failed to separate from booster. Re-entered and burned 26 November 1959.

DISCOVERER VIII

Satellite launched 20 November 1959. Dimensions and payload weights same as DISCOVERER II. Wide elliptical orbit. Inclination to equator, 80.64°. Perigee, 116 miles; apogee, 1038 miles. Capsule separated but was not recovered. Because satellite failed to achieve proper orbit, it was destroyed by g forces on re-entry 7 March 1960.

DISCOVERER IX
Satellite fired 4 February 1960, but did not achieve orbit because of a booster malfunction.

DISCOVERER X
Satellite fired 19 February 1960, but did not achieve orbit because of a booster malfunction.

DISCOVERER XI
Satellite orbited 15 April 1960. Dimensions and payload weight same as DISCOVERER II. Near-circular orbit. Inclination to equator, 80.1°. Perigee, 109 miles; apogee, 380 miles. Capsule was ejected but remained in orbit. Signals ended and booster re-entered 26 April 1960.

DISCOVERER XII
Satellite launched 29 June 1960. Unsuccessful. Carried a re-entry vehicle and recovery capsule, which was instrumented to analyze the events leading to recovery. Objective was to detect reasons for failures of previously attempted recoveries.

DISCOVERER XIII
Satellite launched 10 August 1960. Dimensions and payload same as DISCOVERER II. Near-circular orbit. Inclination to equator, 82.51°. Perigee, 161 miles; apogee, 436 miles. Capsule ejected and retrieved from ocean 11 August 1959 in first recovery of deorbited object. Satellite re-entered atmosphere 14 November 1960.

DISCOVERER XIV
Satellite launched 18 August 1960. Gross weight, 1700 lb. Near-circular orbit. Inclination to equator, 79.6°. Perigee, 116 miles; apogee, 502 miles. Instrumented for radar and optional tracking. Capsule caught in the air as it returned to earth. Satellite re-entered atmosphere 16 September 1960.

DISCOVERER XV
Satellite launched 13 September 1960. Gross weight, 1700 lb. Inclination to equator, 80.93°. Perigee, 130 miles; apogee, 472 miles. Tests included systems evaluation of launching technique, propulsion, communications, orbital performance, advanced

engineering tests, and recovery techniques. Capsule was successfully ejected 25 September 1960 and fell into ocean. Was sighted but lost because of rough seas. Satellite re-entered atmosphere 18 October 1960.

DISCOVERER XVI
Satellite fired 26 October 1960. Gross weight, 2100 lb. Failed to orbit.

DISCOVERER XVII
Satellite launched 12 November 1960. Gross weight, 2100 lb. Inclination to equator 81.8°. Perigee, 116 miles; apogee, 611 miles. Capsule ejected and recovered 14 November 1960, in second air snatch of a deorbited DISCOVERER capsule. Satellite shell re-entered atmosphere 29 December 1960.

DISCOVERER XVIII
Satellite launched 7 December 1960. Gross weight, 2100 lb. Inclination to equator, 80.9°. Perigee, 154 miles; apogee, 459 miles. Capsule ejected and recovered 10 December 1960 after the 48th orbit. Was third air snatch recovery. AGENA B satellite shell still in orbit.

DISCOVERER XIX
Satellite launched 20 December 1960. Gross weight, 2100 lb. Perigee, 128.3 miles; apogee, 390.5 miles. Tests included instrumentation to measure earth's infrared radiation. Carried no recoverable capsule. Satellite re-entered earth's atmosphere 22 January 1961 and burned.

DISCOVERER XX
Satellite launched 17 February 1961. Gross weight, 2450 lb. Perigee 17.7 miles; apogee, 486 miles. Payload included equipment for infrared missile detection which was under test. Malfunction prevented ejection of the 300-lb capsule. Satellite broke into four parts, all still orbiting.

DISCOVERER XXI
Satellite launched 18 February 1961. Gross weight, 2100 lb. Elliptical orbit. Perigee, 149 miles; apogee, 659 miles. Payload was a repeat of DISCOVERER XIX. Did not carry a recovery capsule. Engine was restarted for 1 sec during its first pass over the Pacific Missile Range. Orbital period before restart of engine, 93.8 min. After restart, orbital period was 97.8 min.

DISCOVERER XXII
Satellite launched 30 March 1961. Gross weight, 2100 lb. Perigee, 177 miles; apogee, 486 miles. Payload included instrumentation for cosmic ray research. Failed to remain in orbit, probably because of failure of upper-stage stabilization.

DISCOVERER XXIII
Satellite launched 8 April 1961. Gross weight, 2450 lb. Inclination to the equator, 82.31° (AGENA B); 81.04° (capsule). Elliptical orbit. Perigee, 183 miles; apogee, 388 miles (AGENA B), Perigee, 126 miles; apogee 88.2 miles (capsule). Because of error in attitude and instability in the satellite, capsule could not be returned, as planned, after two days in orbit. After ejection, capsule entered on larger, separate orbit from the satellite shell.

DISCOVERER XXIV
Satellite launched 8 June 1961. Gross weight, 2100 lb. Fell into the ocean shortly after first-stage separation. Nature of payload instrumentation not disclosed.

DISCOVERER XXV
Satellite launched 16 June 1961. Gross weight, 2100 lb. Inclination to equator, 82.11°. Perigee, 139.1 miles; apogee, 251.6 miles. Payload instrumentation included radiation testing equipment and devices to count and measure micrometeoroid impacts. Capsule ejected and retrieved from ocean 18 June 1961 by skin-diving parachutists.

DISCOVERER XXVI
Satellite launched 7 July 1961. Gross weight, 2400 lb. Near-circular orbit. Perigee, 146 miles; apogee, 503 miles. Mission: primarily systems evaluation of AGENA B, particularly of components which had been recently changed; improvement of orbital period control; and ejection and recovery of capsule. After 32 orbits, capsule was ejected and retrieved 9 July 1961 in midair. Was fourth aerial snatch of a DISCOVERER cone, and sixth in the series to be retrieved.

DISCOVERER XXVII
Satellite launched 21 July 1961. Gross weight, 2100 lb. Destroyed 60 sec after launching. No explanation for the failure given by the Air Force.

DISCOVERER XXVIII
Satellite launched 3 August 1961. Failed to orbit. Apparently entire second stage fell into the ocean.

DISCOVERER XXIX
>Satellite lofted 30 August 1961. Payload included bionic experiments. Capsule ejected and recovered 1 September 1961 from ocean north of Hawaii.

DISCOVERER XXX
>Satellite launched 12 September 1961. Capsule ejected and recovered in an aerial snatch as it parachuted toward the ocean north of Hawaii, 14 September 1961. This was the fifth aerial snatch of a DISCOVERER cone.

DISCOVERER XXXI
>Satellite launched 17 September 1961. Capsule failed to separate from the orbiting satellite.

DISCOVERER XXXII
>Satellite lofted 13 October 1961. Experiments in the 300-lb capsule were designed to measure energetic particles which cause changes in samples of yttrium, gold, iron, magnesium, titanium, and nickel; to determine the effectiveness of various metals in shielding from space radiation; and to discover the genetic effect of radiation on seed corn. An experiment to study radio propagation variation through the ionosphere remains in orbit with the AGENA vehicle. The capsule was ejected and recovered by aerial snatch 14 October 1961, sixth aerial recovery in the series.

DISCOVERER XXXIII
>Launch made 23 October 1961. Satellite failed to go into orbit because of a premature shutdown of the AGENA B upper stage following ignition.

DISCOVERER XXXIV
>Satellite lofted 5 November 1961. Length, 25 ft. Weight, 1700 lb. Capsule not recovered because of in-orbit malfunction.

DISCOVERER XXXV
>Launch made 15 November 1961. Capsule recovered on 18th orbit 16 November 1961 by air snatch. Seventh capsule to be so recovered. Three others recovered from sea by divers.

DISCOVERER XXXVI
>Satellite lofted 12 December 1961. Carried piggyback OSCAR. Capsule recovered from ocean by frogmen 16 December 1961.

DISCOVERER XXXVII
Satellite lofted 13 January 1962. Failed to orbit because of mechanical trouble.

DISIL
A paint developed to protect space vehicles from re-entry heat. Consists of a blend of pulverized silicon and other undisclosed ingredients which makes a black paint capable of protecting the metal skins of spacecraft from 3000°F re-entry heat for 45 min, or from 3400° heat for 15 min. Boeing.

DIXIE and PIXIE (Navy)
Code name for project to orbit mice for biomedical research in CALEB satellite.

DME
Distance Measuring Equipment. Second-generation distance-measuring equipment being developed by RCA. Uses solid-state components extensively.

DOCA
See CETIS.

DOD
Department of Defense.

DOG-LEG
Term used for a change in direction of the trajectory of a space vehicle as it is being injected into orbit from a convenient, but not optimum, launching site, e.g., an equatorial orbit can best be attained by equatorial launch, but such an orbit can also be attained from a nonequatorial site by a "dog-leg" trajectory.

DOG-LEG
A lightweight satellite carrying a nuclear charge. Capable of 100,000-mph speed in space. Proposed to the Navy by the Cal-Val Research & Development, spring 1960.

DO-LINER
See GO-DEVIL.

DOLLY (Air Force)
Sounding rocket. Operational. No further details available.

DOLPHIN (Navy)
POLARIS training vehicle without warhead or boosters. Used for simulating launching. Lockheed Missiles and Space.

DOORKNOB (Navy)
    Research rocket developed by Sandia. Two-stage version cap-
    able of lifting 150-lb payload to over 45-mile altitude from
    launch at sea level. Single-stage version: Length, 11.92 ft.
    Weight, 1150 lb. Two-stage version: Length, 18.58 ft. Weight,
    1950 lb.

DOPLOC (Army)
    Doppler radar system.

DOSAAF (U.S.S.R.)
    Initials of the Russian name for "Volunteer Organization for
    Aiding the Army, the Air Force, and the Navy of the Soviet
    Union." Equivalent to a civil defense organization.

DOVE (Navy)
    Military designation, XASM-N-4. An air-to-surface or air-to-
    underwater missile. Similar to ASROC. Primary purpose was
    demolition of underwater targets. Program was initiated April
    1944; canceled February 1955. Eastman Kodak.

D-PAT
    Drum-Programmed Automatic Tester. An automatic electronic
    analyzer used to predict the operational readiness of missile
    systems. Detects and isolates system malfunctions. Hughes.

DRACA or DRACO
    See ALPHA DRACO.

DRAM
    Data-handling system which has an inverted file organization
    for generic-type retrieval.

DRAGON (France)
    High-altitude sounding rocket composed of a BELIER rocket
    with a booster having a 7-ton thrust. Lofts a 40-lb payload to a
    250-mile altitude. Speed, Mach 7.5. Solid propellant. Nonguided.
    Scheduled to be ready by spring 1962. One of a family of inex-
    pensive rockets being developed by Sud-Aviation. See AIGLE,
    BELIER, CENTAURE, ERIDAN, PEGASE.

DRB
    Defence Research Board, Canada.

DRET (Air Force)
    Direct Re-entry Telemetry System. Transmits signals to air-
    borne monitoring equipment through the plasma barrier. Avco/
    Rad.

DROMEDARY
Subsonic missile-launcher aircraft. Capable of high payload and high endurance. Under consideration as possible successor to the B-52 and alternative to the B-70. Probably a turboprop aircraft. Able to stay in air two or more days without refueling. Could be an airborne platform for launching IRBM's. A Rand Corp. proposal to the Air Force.

DRTE
Defence Research Telecommunications Establishment. Part of the Defence Research Board, Canada.

DS-1
See DYNA-SOAR.

DSIF
Deep Space Instrumentation Facilities.

DSN 1,2,3
See GYRODYNE.

DUCK (Air Force)
Military designation, WS-102A/L. Air-to-air bombardment defense missile. Part of the B-58 weapon system. Canceled. Fairchild.

DUMBO (NASA/AEC)
A flying version of the KIWI reactor for the ROVER nuclear rocket engine. Tested successfully.

DUNC (Navy)
Deep Underwater Measuring Device. Radiation-measuring device capable of detecting one atom of radium in 10^{18} molecules of water.

DYNA-MOWS
Manned Orbital Weapon System. Military follow-on to DYNA-SOAR. To be larger than DYNA-SOAR. Boeing directing two subcontract teams on competitive design.

DYNA-SOAR (Air Force)
Military designation, WS-464L. A manned space glider program which entails the development of a manned glider capable of suborbital and orbital flight. First suborbital flights are scheduled for 1963-64 using a TITAN II booster. Orbital flights in 1964-65 will be boosted by SATURN. Scale models will be tested using

SCOUT. The Curtiss-Wright HTV-2 was used in early model testing in 1959. DYNA-SOAR studies began as early as 1954, but only in November 1959 was Boeing selected to develop the glider and Martin the booster. The DYNA-SOAR program as presently planned is a logical outgrowth of the X-15 program. The DYNA-SOAR glider is a snub-nosed, delta-shaped craft. Delta is tailless and low-winged with low-aspect ratio, and with vertical tail surfaces. Over-all length of present glider, about 35 ft. Weight, about 10,000 lb (suborbital) and 20,000 lb (orbital). Fins will be added to the TITAN II booster for stability. First-stage thrust about 430,000 lb; second-stage thrust, about 100,000 lb. Estimated total cost for DYNA-SOAR, about $800 million. Allocation for FY 1962, $185.8 million. Boeing, glider and systems integration; Martin, booster; Aerojet-General, propulsion; Minneapolis-Honeywell, guidance; RCA, communications.

E

E-1
400,000-lb-thrust engine, liquid propellant. Under R&D. Rocketdyne.

E-3C
See POGO-HI.

E-4
Military designation for Air Force MIDAS.

EAGLE (Navy)
Military designation, XAAM-N-10. Air-to-air missile. Designed primarily to be launched from the Missileer aircraft. Range, estimated at 100 nm. Speed, more than Mach 3. Canceled in its present form. May be revived using the FY 1963 state of the art. Bendix.

EARLY SPRING (Navy)
Antisatellite missile program. Ships and platforms to launch Navy space mines. These would be vertical interceptor satellites containing optical scanning system capable of discriminating between stars, known friendly vehicles, and suspect orbiting objects; identifying the target; and relaying data to the ground for positive control of destruction, if desired.

ECHO I (NASA)
Pliant, passive communications satellite, launched from Cape Canaveral 12 August 1960. Diameter, $26\frac{1}{2}$ in. before inflation;

100 ft, after inflation. Gross weight, 240 lb; net weight, 136 lb. Lofted by THOR-DELTA booster into an equatorial, circular orbit. Perigee, 945 miles; apogee, 1049 miles. Langley Research Center, prime; Schjeldahl, inflatable structures.

ECHO II (NASA)
Rigid, passive communications satellite, 135 ft in diameter. Weight, 600 lb. Scheduled for launch into 650-nm circular orbit from Pacific Missile Range in first half of 1962. THOR AGENA B booster. Balloon to be made of laminated aluminum foil 0.0002 in. thick, sandwiching a layer of Mylar plastic 0.00035 in. thick. To be painted black inside and dotted white outside to distribute heat and maintain uniformity of temperature within the balloon. Schjeldahl has contract to build nine satellites. Cost, $400,000.

EDDS (Air Force)
Early Docking Demonstration System. Study contracts to Lockheed Missiles and Space and STL, 1961.

EDSAC
Cambridge University Mathematical Laboratory Computer.

EGAD
An electronegative gas detector which detects and measures gas concentrations as low as one part in 10 million. A thermoelectric refrigerator (0.33 in.^{3}) handles excess water vapor in the gas samples. Westinghouse.

EGO (NASA)
Eccentric Geophysical Observatory, part of OGO program. Designation, S-49. Designed to measure atmosphere, ionosphere, earth's magnetic field, in eccentric orbit with 170-mile perigee, 10,000-mile apogee. Payload weight, 1000 lb, including 150-lb experiments. Design has growth potential to 1500 lb, including 300-lb piggyback satellite. Size of EGO: length, 6 ft; 3 ft square. To carry 19 experiments; however, modular compartments and other space in EGO body will later be able to carry as many as 50 experiments on one mission. To be lofted by ATLAS-AGENA B, January 1963. Later satellites to use THOR-AGENA (POGO) and CENTAUR boosters. Backup to S-49 has launch date set tentatively from January to September 1963. STL has contract for three flight spacecraft within 3 years. NASA plans to loft one or two OGO satellites (EGO and POGO) a year for the next 5 years. OGO program expected to cost $50 to $100 million in 8 years.

ELR (Great Britain)
Experimental Launching Round. Early attempt by Royal Navy and Armstrong Whitworth to develop a useful surface-to-air missile. Diameter, 4 in. Liquid propellant. Launched vertically from shipboard and oriented in flight by a steam jet produced by a steam generator which was mounted internally.

EMERAUDE (France)
See TOPAZE.

EMSS (Air Force)
Emergency Mission Support System. Mission to determine what kind of avionic systems are needed for the 1965-75 time period. Directed by Cambridge Research Laboratories and Electronic Systems Division. Study contracts to Airborne Instruments Laboratory, Alpha, Avco, Garrett, Gilfillan, Hazeltine, HRB-Singer, ITT Labs, Lockheed, Maxson, North American, and RCA.

ENDORSE
A study of the effects of controlled isolation upon persons performing some 30 different kinds of tasks.

ENEA
European Nuclear Energy Agency. Under auspices of ENEA, 13 European countries are working together in Belgium in the Eurochemic project on irradiated fuels. Twelve countries are working on the Halden project in Norway, and 12 are working on the Dragon reactor program in England.

ENERGETIC PARTICLES SATELLITE
See EXPLORER XII.

ENTAC (France - Army)
ENgin Téléguide Anti-Char. Surface-to-surface antitank missile. Missile in-flight weight, 27 lb; missile and launcher weight, 37 lb. Range, 1 mile. Speed, 180 mph. Solid propellant. Wire guidance. HE warhead. Turned over to Nord by DEFA. Bought by U.S. Army, replacing the SS-10. Also bought by Belgian Army. In production. Joint Nord Aviation and DEFA development.

ENVOY
Three-stage, solid-propellant vehicle capable of lofting a 50-lb payload to the moon or putting a 230-lb payload into a 300-mile orbit. Vehicle weight, 17,000 lb. Proposed by Grand Central Rocket, 1959, as alternative to SCOUT.

ERIDAN (France)
> Single-stage, high-altitude sounding rocket identical to AIGLE, except that a supplementary booster is attached. Will loft a 220-lb payload to a 220-mile altitude, or a 660-lb payload to a 100-mile altitude. Solid propellant. One of a family of inexpensive rockets being developed by Sud-Aviation. See AIGLE, BELIER, CENTAURE, DRAGON, PEGASE.

ERIKA (Sweden)
> Ship-to-underwater, antisubmarine rocket. Length, 6.7 ft. Weight, 550 lb. HE warhead. Longer range than the AURORA, more than 1 mile at 170-ft depth. Under evaluation. Bofors.

ERO
> European Research Office (Dept. of Army) Frankfurt, Germany. In 1956 began enlisting European scientists in U.S. Army-supported unclassified research. Army regulations 70-40.

ESAR (Air Force/ARPA)
> Electronically Steerable Array Radar. Capable of focusing its electric eye simultaneously on both space vehicles and aircraft at lower altitude. Part of DEFENDER. Bendix.

ESD
> See AFESD.

ESLO
> European Space Launcher Organization. An organization set up to determine what boosters should be used for European space programs. The British BLUE STREAK has been adopted for the first stage of a large booster. The second stage is the French VERONIQUE. Third stage will be a West German rocket.

ESRO
> European Space Research Organization. An organization of 11 states of western Europe, initiated in 1960 for the purpose of conducting space research. Group was formed through the initiative of Britain and France because neither country felt itself able to afford the research alone. Member nations make contributions in proportion to their national income. See COPERS, ESLO.

EURATOM
> Communauté européene de l'énergie atomique (EURopean community for ATOMic energy). A European organization formed to advance the peaceful uses of atomic energy.

EX-8 (Navy)
    Antisubmarine, rocket-powered torpedo missile. In R&D. Aero-
    jet-General/Bendix.

EXCELSIOR (Air Force)
    Project to test physiological actions and reactions of a man in
    an escape situation at 100,000-ft altitude by means of a high-
    altitude balloon flight and parachute jump. Performed at Hollo-
    man AFB, November 1959, under the sponsorship of Aerospace
    Medical Laboratory, Dayton, Ohio. Capt. Joseph Kittinger para-
    chuted from the balloon used in the MANHIGH program.

EXOS (Army)
    Three-stage vehicle designed to loft 50-lb payload to 300- to
    450-mile altitude. Under development for the Army by the
    University of Michigan. Reportedly, the Navy is interested.

EXPLORER
    Lockheed proposal to launch a series of probes into the upper
    atmosphere using the X-17 re-entry test vehicle, 1957. Not
    adopted.

EXPLORER (Army/NASA)
    Name given to a series of early research satellites launched
    by various boosters (JUPITER C, THOR-ABLE, JUNO II, SCOUT,
    THOR-DELTA) starting in January 1958. At first an ABMA pro-
    gram, later a NASA series. See EXPLORERS I – XIII.

EXPLORER I (ABMA)
    Research satellite developed from concepts of Project ORBITER.
    Gross weight, 30.8 lb; net weight, 18.13 lb. Launched from Cape
    Canaveral 31 January 1958 by JUPITER C booster into an
    elliptical orbit. Inclination to equator, 33.34°. Perigee, 224 miles;
    apogee, 1573 miles. Payload instrumentation measured internal
    temperature, erosion, cosmic rays, and micrometeoroid im-
    pacts. Discovered the inner radiation belt around the earth,
    identified by Dr. J. A. Van Allen. Satellite's estimated lifetime,
    3 to 5 years. Designed by University of Iowa and Jet Propulsion
    Laboratory.

EXPLORER II (ABMA)
    Research satellite launched 5 March 1958 by JUPITER C booster.
    Total payload weight, 31.5 lb. Failed to orbit because last stage
    of booster did not ignite. Satellite fell into the ocean 3000 miles
    from Cape Canaveral.

EXPLORER III (ABMA)
Research satellite launched 26 March 1958 by JUPITER C booster. Length, 80 in. Diameter, 6 in. Gross weight, 31 lb. Net weight, 18.56 lb. Wide elliptical orbit. Inclination to equator, 33.4°. Perigee, 121 miles; apogee, 1746 miles. Instrumentation recorded data on the radiation belt discovered by EXPLORER I, on micrometeoroid impacts, and on internal and external temperature of the satellite. Because of close perigee, re-entered atmosphere 28 June 1958.

EXPLORER IV (ABMA)
Research satellite launched 26 July 1958 by JUPITER C booster. Length, 80.4 in. Diameter, 6.25 in. Gross weight of payload, 38.43 lb; net weight, 25.8 lb. Wide elliptical orbit, with 50.29° inclination to equator. Perigee, 163 miles; apogee, 1380 miles. Carried lead-shielded Geiger counter and two scintillation counters. Instrumentation recorded data on radiation belts. Re-entered atmosphere 23 October 1959.

EXPLORER V (ABMA)
Research satellite launched 24 August 1958 by JUPITER C booster. Failed to orbit because of a collision between parts of the booster and the instrument compartment which carried the three high-speed stages. Carried 38.4-lb payload.

EXPLORER VI (NASA)
Also called PADDLEWHEEL, ABLE III. Satellite with four solar cell "paddlewheels" launched 7 August 1959 from Cape Canaveral by THOR-ABLE III booster. Length, 2.42 ft. Diameter, 26 in; with paddles extended, approximately 8 ft. Gross weight of payload, 192 lb; net weight, 142 lb. Extreme elliptical orbit. Inclination to the equator, 47.85°. Perigee, 156 miles; apogee, 26,357 miles. Experiments included measuring three specific radiation levels of earth radiation belt, mapping the earth's magnetic field, and detecting micrometeoroids. Satellite made first TV transmission of earth's cloud cover.

EXPLORER VII (NASA)
Also called RADIATION SATELLITE. Launched from Cape Canaveral 13 October 1959 by JUNO II booster. Length, 29 in. Diameter, 30 in. Gross weight of payload, 91.5 lb; net weight, 70 lb. Near-circular orbit. Inclination to equator, 50.45°. Perigee, 346.6 miles; apogee, 678 miles. Satellite discovered solar X rays and storms of solar radiation.

EXPLORER VIII (NASA)
Also called RADIATION LABORATORY. NASA designation, P-46. Geophysical probe launched from Cape Canaveral 3 November

1960, JUNO II booster. Length, 30 in. Diameter, 30 in. Payload
net weight, 90.14 lb. Wide elliptical orbit. Inclination to equator,
49.9°. Perigee, 258 miles; apogee, 1423 miles. Instrumentation
measured electron temperatures and electron distribution in the
ionosphere, and recorded energy of striking micrometeoroids.
Estimated lifetime, 20 to 50 years.

EXPLORER IX (NASA)
Also called POLKADOT BALLOON, BABY ECHO. Designation,
S-56. Twelve-foot balloon satellite made of aluminum-coated
plastic, lofted by SCOUT booster from Wallops Island, Va., 16
February 1961. Payload gross weight, 80 lb; net weight, 65 lb.
Wide elliptical orbit. Inclination to equator, 38.63°. Apogee,
1605 miles; perigee, 395 miles. Estimated lifetime, 1 year.

EXPLORER X (NASA)
Also called PLASMA PROBE. Designation, P-14. Scientific
satellite lofted 25 March 1961 from Cape Canaveral by THOR-
DELTA booster. Purpose: to study magnetic fields and solar
winds. Length, 52 in. Diameter, 19.36 in. (base), 13 in. (top).
Payload gross weight, 78 lb. Extreme elliptical orbit. Apogee,
112,500 miles; perigee, 110 miles. Varian.

EXPLORER XI (NASA)
Also called GAMMA RAY TELESCOPE. Scientific satellite lofted
27 April 1961 from Cape Canaveral by JUNO II booster. Length,
20.5 in. (base), 23.5 in. (box). Diameter, 6 in. (base), 12 in. (box).
Payload gross weight, 94.8 lb; net weight, 82 lb. Wide elliptical
orbit. Inclination to equator, 28.8°. Perigee, 304 miles; apogee,
1113.2 miles. Instrumentation measured gamma emissions of
Crab Nebula and made analysis of interstellar cosmic rays.
Satellite tumbling end over end at controlled rate of about 10
rpm. Estimated life in orbit, 13 months. Designed by M.I.T.

EXPLORER XII (NASA)
Also called ENERGETIC PARTICLES SATELLITE. Designation,
S-3. Scientific satellite lofted 15 August 1961 from Cape Canaveral by THOR-DELTA booster. Weight, 82 lb. Octagonal shape.
Wide elliptical orbit. Perigee, 172 miles; apogee, 54,000 miles.
Major purpose of flight: study of the interaction between solar
winds and the earth's magnetic field and other magnetic fields
in space. Also instrumentation gathered detailed information on
the Van Allen radiation belts.

EXPLORER XIII (NASA)
Satellite lofted 25 August 1961, SCOUT booster. Weight, 190 lb.
Five micrometeoroid detectors, designed by NASA's Langley

and Lewis Research Centers and Goddard Space Flight Center, were wrapped around the booster's fourth stage ALTAIR rocket. Satellite achieved orbit which then decayed to a perigee of 70 to 75 miles. After 2 days in orbit, EXPLORER XIII re-entered the atmosphere.

F

F-1 (NASA)
Single-chamber rocket engine designed to produce 1.5 million-lb thrust. In R&D. Qualification tests scheduled for 1965. Probably available by 1966. Under NOVA program. North American.

F-42 (Army)
Fireball to deliver Napalm on target. Army Chemical Corps.

F-89D SCORPION
See FFAR.

F-101B
See VOODOO.

F-104
See SUPER STARFIGHTER.

F-105A
See THUNDERCHIEF.

F-106
See DELTA DART.

FAA
Federal Aviation Authority.

FABMIDS (Army)
Mobile antimissile defense system. Expected to interest NATO. Feasibility study contracts to G.D. – Astronautics, Hughes, Martin, G.E., Raytheon, and Sylvania, October 1960.

FAGMS-S
See SERGEANT.

FAIREY A-T (Great Britain – Army)
Surface-to-surface antitank missile. Reportedly, radar-guided. May be comparable to Pye P. V. In study phase. Fairey Engineering.

FALCON (Air Force)
Military designations, GAR-1, -2, -3A, -4A, -9, -11; XAAM-A-2.
Air-to-air guided missile. Guidance: radar-homing, GAR-1; IR
homing, GAR-2. Length, 6 to $6\frac{1}{2}$ ft. Weight, 110 to 120 lb. Solid
propellant. Speed, Mach 2. Range, 5 miles. Thrust, 6000 lb.
Warhead: HE for GAR-1 through -4; GAR-11, nuclear. GAR-1
and -2 operational but discontinued; GAR-3, -4, -11 operational.
GAR-9, in production. GAR-3 and -4 known as SUPER FALCONS.
Engines: see XM18; XM18E1; XM-18E3; XM46; M58E1; M58E2.
Hughes, prime; Hughes, guidance; Thiokol, propulsion.

FARSIDE (Air Force)
Four-stage, high-altitude research rocket. Solid propellant.
Launched from balloon 19 miles above earth. Altitude, nearly
3000 miles. Program purpose: to acquire scientific data at
extreme distances from earth. Six firings were made during
September and October 1957. Length, 24 ft. Weight, loaded,
1970 lb; payload, 3 to 5 lb. Speed, 18,000 mph. Composed of
booster of four RECRUITS; second stage, one RECRUIT; third
stage, four ARROW 2's; fourth stage, one ARROW 2. Ford-
Aeronutronic.

FAST (Navy)
Fast Automatic Shuttle Transfer. System of resupplying ships
at sea while steaming toward their objective. Capable of handling
not only all types of general stores, but also ordnance, including
missiles. PneumoDynamics.

FBM
Fleet Ballistic Missile. See POLARIS.

FBM-1
See POLARIS.

FEAR
See MIGHTY MOUSE.

FELIX (Army/Air Force)
Basically a vertical bomb like the RAZON (VB-3). After launch,
its target-seeking infrared device would home in on the heat
radiation given off by the target.

FELIX I (Brazil)
A biological sounding rocket made from a modified SS-10 rocket
launched vertically. Altitude, 112 miles. Maximum velocity,
4470 mph. Part of Brazil's research rocket program.

FERO
    Far East Research Office (Dept. of Army). Camp Zama, near
    Tokyo, Japan. Started 1959. At present, largely confined to un-
    classified medical research.

FFAR (Air Force)
    Folding Fin Aircraft Rocket. Carried in clusters in F-89D
    SCORPION wing-tip launching pods.

FID
    Fédération Internationale de Documentation (International Fed-
    eration of Documentation). Part of UNESCO.

FIESELER FI-103
    See V-1

FIGHTER
    Improvement of the effectiveness of humans in a man-machine
    system is aim of this project. See ENDORSE.

FINGERPRINT (Air Force)
    Guidance system proposed for use in SLAM. Under considera-
    tion by the Air Force. Developed by Chance-Vought.

FIRE
    One of a planned family of tactical short-range missiles. Liquid
    propellant. Purpose: to crush enemy defense in front of ad-
    vancing troops. Fired for Army. Developed by Chance-Vought.
    Company-funded.

FIREBEE (Air Force)
    Military designations: XQ-2A,B,C; Q-2A,B,C (Air Force); KDA-1
    (Navy); XM-21 (Army). Target missile. One of the first custom-
    made drones. Special devices can be attached to the missile for
    simulating varied target characteristics. Ryan Aeronautical.

FIREBEE (Air Force)
    Radar-guided target drone which duplicates performance and
    radar profiles of supersonic high-altitude bombers. Records
    hits and misses of missiles fired at it. Radioplane.

FIREBIRD (Air Force)
    Air-to-air missile. Solid propellant. Radar-homing. Four mis-
    siles could be launched simultaneously from underwing racks.
    First Air Force air-to-air missile. Produced in production
    quantity in 1952. Ryan Aeronautical.

FIREFLASH (Great Britain - RAF)
Air-to-air missile. Solid propellant. Speed, Mach 2+. Deployed with RAF squadrons for training purposes. HE payload. Length, 7 ft 5 in. Launch weight, 300 lb. Beam-riding. Has been replaced by the FIRESTREAK. Fairey Aviation.

FIREFLY
Missile similar to small jet aircraft with rotor blades in tail. Developed by Solar Aircraft for peaceful uses, e.g., fighting forest fires, nuclear fires; aiding downed planes, particularly where there is danger of fire.

FIREFLY (Army/Navy/Air Force/NASA)
Former name for ANNA.

FIREFLY (ARPA/Air Force CRL)
Program of 33 rocket firings from Eglin AFB. Purpose: basic research of upper atmosphere, production of artificial iono-spheres to test reflections of radio signals, and study of physical and chemical reactions in upper atmosphere resulting from passage of an object. Cost, about $1 million. Aerojet-General.

FIREPOWER (Army)
Surface-to-surface missile. Purpose: improvement of the effec-tiveness of the assault power of tanks. In R & D.

FIRESTREAK (Great Britain - RAF)
Air-to-air missile. Solid propellant. Length, 10.5 ft. Launch weight, 3000 lb. Infrared guidance. HE payload. Speed, Mach 2.3. Range, 4.5 miles. Operational. DeHavilland Propellers.

FIRETRAC (Navy)
System which measures relative trajectory, velocity, and miss-distance of a missile fired at a target drone. Developed by Aerojet-General, Atlantic Div.

FISHBOWL (Navy)
An active ocean surveillance project. Part of TRIDENT. In R&D.

FLAG
See CSAR.

FLEDGLING (NASA)
A plan considered in 1958-59 to research the composition of the sun and extreme upper atmosphere by orbiting 12 satellites at 5000- to 15,000-mile distance. Booster composed of VANGUARD first stage and advanced SERGEANT second stage.

FLEX WING
> A small flexible-wing aircraft developed by Ryan. The wing, mylar-bonded to nylon, has a span of 40 ft. Lifting power is derived from flow of air which inflates the wing. 180-hp Lycoming engine. Total weight, 1500 lb. Possible uses: as auxiliary wing for use in take-off; to reduce landing speeds; booster recovering; emergency wings; re-entry gliding; helicopter tow of payloads; and reconnaissance drones.

FLIGHT RESEARCH CENTER (NASA)
> Located at Edwards, Calif. Center for aeronautical research and test, including the X-15.

FLIP (ONR/Marine Phys. Lab/Scripps Institute of Oceanography)
> Oceanographic research vessel with 30-story deep tubular hull. Designed primarily to make breakthrough in underwater submarine detection. Length, 355 ft from top to bottom. Another oceanographic research vessel is the SEA QUEST (Lockheed). See also SPAR.

FLORA (Sweden)
> Antisubmarine rocket. Longer range than AURORA. Bofors.

FLUTTER DART (Great Britain)
> Solid-propellant research rocket used to investigate the aerodynamic characteristics of supersonic vehicles. Armstrong-Whitworth Aircraft.

FLUXLOX
> Ferrite core memory device under development by Burroughs Corp.

FLYING SPY (Army)
> Surveillance drone. Solid-propellant booster. Parachute recovery system. Developed by Rheem which was later bought by Aerojet-General, now the producer of the advanced SD-2.

FORTRAN
> A computer system.

FPTS
> Forward Propagation Tropospheric Scatter. A radio communication method, part of the WHITE ALICE system which makes use of remote sites in the far north as radio transmitter and receiver stations.

FRAM (Navy)
Fleet Rehabilitation And Modernization.

FRANSPANO
French-Spanish dictionary compiled by machine translation. University of Mexico.

FREEDOM 7
Name given MERCURY capsule which carried Cdr. Alan Shepard into a ballistic trajectory.

FRESCAN (Navy)
Lightweight, all-weather radar. Used in missile cruisers and destroyers armed with TALOS, TERRIER, and TARTAR missiles. Hughes.

FRIDA, GERDA, ADAM (Sweden – Air Force)
Air-to-air missiles, operational on Saab, Lansen, and Draken aircraft. Frida has 2624 ft/sec velocity at 32,000 ft. Burning time, 0.9 sec. GERDA is improved version of FRIDA. ADAM is an air-to-surface version of FRIDA. Length, about 6 ft. Weight, 90.4 lb. HE warhead. Used for strafing. Bofors A. B.

FROST (Air Force)
Project to provide methods of storing food on space trips. G. E.

FX
Air Force tactical fighter plane.

G

GAM
Guided Aircraft Missile.

GAM-63
See RASCAL.

GAM-67
See CROSSBOW.

GAM-72
See QUAIL.

GAM-77
See HOUND DOG.

GAM-79
    See WHITE LANCE.

GAM-83
    See BULLPUP.

GAM-87A
    See SKYBOLT.

GAMBIT
    Martin proposal to orbit two or three men at 200- to 300-mile
    altitude to study effects of weightlessness and other environ-
    mental conditions of humans. Company-funded.

GAMMA RAY TELESCOPE
    See EXPLORER XI.

GANDER (Air Force)
    Air-to-air or air-to-surface plastic-constructed missile. Sub-
    sonic. Payload, 1 MT. Range, 2000 miles. Similar to BULL-
    GOOSE or GOOSE, but with nuclear warhead. Fairchild.

GAPA (Air Force)
    Ground to Air Pilotless Aircraft. Two-stage, antiaircraft
    missile. Speeds to 1500 mph. Altitude, 6000 to 8000 ft. GAPA
    research used in developing later BOMARC. Discontinued in
    1949. Boeing.

GAR
    Guided Aircraft Rocket.

GAR-1, -2, -3, -4, -9, -11
    Also GAR-1D and GAR-2A. See FALCON.

GAR-8
    See SIDEWINDER.

GARGOYLE (Navy)
    Liquid-propellant, air-to-surface missile. Radio-controlled and
    gyro-stabilized. Launched from carrier-based planes. Project
    initiated in 1943 as a powered glide bomb by Bureau of Aero-
    nautics (Navy) after comparative evaluation of the GLOMB,
    GORGON, and first design of the GARGOYLE. Later designs of
    the GARGOYLE were based on need for a missile to pierce
    armored decks of ships. After World War II, program was re-
    turned to developmental status. Later was used to test missile
    components at Mojave Naval Air Station. McDonnell Aircraft.

GASLIGHT (Army)
Program using JUPITER (some POLARIS and THOR firings) in a study of re-entry radiation from IRBM's. Begun in 1958.

GASP
Another name for ORION (PUT-PUT).

GASP (Navy-ONR)
Study of launching sites and launching methods for space vehicles. Classified. J. H. Pomeroy.

GAT
Georgetown Automatic Translation. A computer translation system. Georgetown University.

GB
Glide Bomb.

GDD-1
Miniature drone. Purpose: to provide meteorological air pollution data for U.S. Weather Bureau and Public Health Service. Rocketdyne.

GEERS
Groupe Européen d'Etudes pour les Recherches Spatiales. Sixteen-state conference, consisting of Austria, Belgium, Denmark, France, German Fed. Rep., Great Britain, Greece, Iceland, Ireland, Italy, Luxembourg, the Netherlands, Norway, Sweden, Turkey, and Spain.

GEESE
General Electric Electronic System Evaluator. A hybrid computer system in which a great variety of electronic signals can be selectively simulated.

GEM
Ground Effect Machine. Generic term. See also SES (Surface Effect Ship).

GEM (DOD - ONR)
Fifteen-ton Ground Effect Machine. Purpose: to replace amphibious landing craft and to be capable of crossing a beach and unloading further inland. Similar to Army LOTS. Avro/Borg-Warner.

GEMINI (NASA)
Also called MERCURY II. Enlarged, more sophisticated version of MERCURY capsule. Size: 1 ft larger than MERCURY (75-in. base) and lengthened proportionately to enlarge cabin volume by 50%. Weight, 4000 lb during first flights; 6000 lb for rendezvous missions. TITAN II booster, with second stage uprated from 15,500 mph to 17,500 mph. To seat two astronauts side by side. Several unmanned ballistic flights set for 1963; manned flights for 1 day, 3 days, and 1 week to follow. Orbital rendezvous scheduled for 1964.

GENIE (Air Force)
Air-to-air supersonic missile. Length, 8 ft, 6 in. Range, 6 miles. Weight, about 800 lb. Operational. SUPER GENIE to be guided, with more powerful propulsion and greater range. Also known as DING-DONG and HIGH CARD. Engine: See XM29. Douglas, prime; Aerojet-General, propulsion.

GEORGE C. MARSHALL SPACE FLIGHT CENTER (NASA)
Located at Huntsville, Ala. This center has the principal responsibility of developing big boosters for the National Space Program. The principal boosters at present are SATURN and NOVA.

GERDA
See FRIDA (Sweden).

GESOC
General Electric Satellite Orbit Control. Device consisting of a computer to advise of orbital corrections required, and a solid-rocket gas generator to perform the orders. Infrared sensor determines orbit parameters.

GFE
Government Furnished Equipment.

GHOST
Global HOrizontal Sounding Technique. Tele-Dynamics in-house development to demonstrate feasibility of avionics systems of low mass and low physical density. Systems attached to pillow-shaped balloon below super-pressure balloons. Purpose: to gather meteorological data in stratosphere and troposphere. American Bosch Arma, Tele-Dynamics Div.

GIMLET (Navy)

Air-to-surface missile. Solid propellant. Unguided. Range, 1000 miles. Outgrowth of FEAR. Used by attack-type aircraft. Limited production. Inactive. NOTS, prime; Hunter-Douglas.

GIMRADA (Army)

Geodetic Intelligence Mapping Research And Development Agency, Fort Belvoir, Va. Purpose of agency: to write payload specifications by collecting and plotting reconnaissance data.

GIPSE

Gravity-Independent PhotoSynthetic gas Exchanger. A system which uses algae to generate oxygen and remove carbon dioxide from the passenger compartment of a space capsule. Also checks oxygen supply system during zero gravity. Test to be made in an ATLAS E pod from AMR.

GLADEYE

Container for dropping large quantitites of World War II bombs from a jet aircraft. Way of using existing stocks in a more effective way.

GLIDE

Theoretical concept of missile gliding rather than following a pure ballistic path. Missile would climb to 50-mile height, approximately, then glide horizontally toward target at 10,000-plus mph. Under study.

GLIMB (Navy)

An antisubmarine pilotless aircraft. Would carry 650-lb bomb and have a radio control gear with a flare for direct sighting to guide it to the target. To be operated from a blimp. Never used.

GLIPAR (ARPA)

Guide Line Identification Program for Antimissile Research. Nine-month study in ballistic missile defense field. Investigated antigravity and antimatter. Cost, $ 1.6 million. Part of DE-FENDER.

GLOBEQUICK (Air Force)

Control system to provide rapid communications between USAF headquarters and major operation commands. Air Force.

GLOMB (Navy/Army)

Military designations: XLNT-1,2; LBE-1; LBP-1; SRB; TO-6. Glide bomb device, Navy development. Program begun in 1941, but discontinued in 1944 since course of the war had changed. Bomb was capable of carrying 3000 gallons of gasoline or 18,000 lb of bombs. GLOMB weight, 7320 lb with 4000-lb payload. Speed, 300 mph. Pratt-Read.

GM
Guided Missile.

GNAT (Navy)
Surface-to-air rocket. In R&D. Naval Ordnance Lab.

GNOME (AEC)
Project to discover feasibility of storing heat from nuclear blasts underground for use in generating electricity. A 1200-ft shaft driven down into a salt bed in sourtheastern New Mexico with a tunnel parallel to the surface. Plan: to set off a 10-kiloton bomb at the end of the tunnel. Completion of tunnel late 1961. Recently GNOME program has been expanded to include tests for developing means of detecting underground nuclear blasts. See also VELA.

GOBI (Air Force)
Air-to-surface missile. Program initiated April 1946; canceled May 1947.

GODDARD SPACE FLIGHT CENTER (NASA)
Located at Greenbelt, Md. Responsible for scientific research in space, manned flight, including MERCURY, communication, and meteorological satellites.

GO-DEVIL
Ground-support amphibious vehicle capable of moving over obstructions by means of levered wheels. Being evaluated by services. Capable of walking up and over a 6-ft vertical bank or traversing a 45-degree slope with the chassis remaining on even keel. Could be used as a carrier for personnel, weapons, or cargo. Wagner Tractor.

GODIVA II (AEC)
Reactor with bare uranium core developed by Los Alamos Scientific Laboratory.

GOLDEN RAM (Air Force)
Program to tighten operational launch procedures and eliminate problems causing partial failures. ATLAS missile launched successfully by SAC, Vandenberg AFB, 16 December 1960.

GOLEM I (U.S.S.R. - Navy)
Underwater-to-surface missile. Length, 53.8 ft. Weight, 33,125
lb. Altitude, 125 miles. Range, 395 miles at Mach 7. Liquid-
fueled, radio inertial guided. 2000-lb HE or nuclear warhead.
Two or three missiles can be towed behind a submerged submarine. Missile stands upright in water by flooding the ballast
chamber in the tail. Developed from German World War II plans
for a seagoing V-2. Operational.

GOLEM II (U.S.S.R. - Navy)
Underwater-to-surface missile. Advanced version of GOLEM I,
with improvements adopted from the Army T-2. Length, 56.9 ft.
Weight, 74,800 lb. Range, 1240 miles at Mach 11.4. Liquid propellant. 1430-lb HE or nuclear warhead. Programed guidance.
Three missiles towed behind submerged submarine. Missile
stands upright in water by flooding the ballast chamber in the
tail. Operational.

GOLEM III (U.S.S.R. - Navy)
Underwater-to-air and surface-to-air missile. Length, 17.2 ft.
Weight, 4625 lb. Range, 7.5 miles at Mach 2.5. Solid propellant.
Infrared guidance. Launched from tube by compressed air. Land
version in development stage for Soviet Army. Antiaircraft protection for both submarines and surface ships. Operational on
surface ships.

GOLEM IV (U.S.S.R. - Navy)
Surface-to-air missile. New. Radar-guided and solid-fueled.
May be converted to submarine use also. Range, 45 miles.
Operational on surface ships.

GOOSE (Air Force)
Air-to-surface and air-to-air missile used as decoy by SAC to
protect bomber crews over enemy territory. Engine: XM38.
Canceled December 1958 after an expenditure of $78.5 million.
Fairchild.

GORGON (Navy)
Air-to-surface missile developed to test feasibility of the ramjet engine. Canceled 19 December 1949. See PLOVER. Martin.

GOSS
Ground Operational Support System.

GPSSM (Army)
General Purpose Surface-to-Surface Missile. Advanced tactical missile. Surface to surface. Under study. No contracts announced.

GRAPEFRUIT
Name given VANGUARD I by Russians.

GREB (Navy/ARPA/Air Force)
Galactic Radiation Experiment Background. Also called SUNRAY. Solar radiation study. Launched 22 June 1960 as TRANSIT IIA piggyback. Booster, THOR ABLESTAR. Transmitting first continuous measurements of solar activity in ultraviolet and X-ray radiation. Second GREB piggyback on TRANSIT doubled with INJUN. NRL.

GREBE (Navy)
Surface-to-surface missile program initiated August 1944; canceled July 1951.

GREBEL (Navy)
Air-to-surface missile. In R&D.

GREEN QUAIL
See QUAIL.

GRIFFON (France)
Supersonic experimental interceptor. Development of turbojet-ramjet propulsion by Nord Aviation. Funded by U.S. Air Force.

GRISA
See CETIS.

GSE
Ground Support Equipment. See also AGE.

GSFC
See GODDARD SPACE FLIGHT CENTER.

GT
Glide Torpedo.

GVAI (U.S.S.R. - Army)
Barrage rocket fired in salvos from multiple-tube launcher. May be phased out for advanced versions.

GYRODYNE (Navy)
   Military designation, DSN 1,2,3. A drone helicopter for use in
   attacking submarines by means of homing torpedoes. Uses
   radar/sonar guidance system.

H

H-1 (NASA)
   188,000-lb-thrust SATURN engine under development by Rocketdyne.

HARDTACK
   Space research project using ARGO rockets to measure corpus-
   cular radiation caused by nuclear bombs fired at Eniwetok,
   August 1958.

HARE (Air Force)
   Spacecraft designed for high-altitude, recombination-energy
   propulsion. Vehicle would fly at 60-mile altitude where it would
   be fueled continuously by "free" oxygen, changed to atomic form
   by ultraviolet radiation. Velocity, about 400 miles/sec. Study
   made by Rennselaer Polytechnic Institute.

HARE IN THE AIR (Air Force)
   Biomedical experiment in which a rabbit is launched by BLUE
   SCOUT booster to determine amount of radiation to which the
   animal is subjected. Most advanced of ten biomedical experi-
   ments planned for flight during FY 1962 by Aerospace Medical
   Laboratory at Wright-Patterson AFB.

HARP (Australia)
   High Altitude Research Project. Rocket launched by balloon.
   Balloon built of plastic reinforced with glass fiber. Solid propellant. Weight of launching balloon, 520 lb; 100 lb empty. Capable of lifting 750 lb at 1000 ft/min. Firing done by radio command. Altitude, 57 miles. Developed by Fairey Aviation, Aus-
   tralasia; sponsored by the Australian Weapons Research Establishment.

HARPY (Army)
   Small surface-to-air missile to be fired from bazooka against
   low-flying aircraft. Electro-optical guidance. Audiosonics.

HAS
High Altitude Sample. Multistage test vehicle composed of seven VIPER engines plus one LANCE or NIKE. Capable of carrying a 60-lb payload to a 500-mile altitude. Cost, $30,000. Sandia.

HASP (Navy)
High Altitude Sounding Program. Similar to Army's LOKI missile. A research weather rocket capable of attaining a 35-mile altitude. Operational since 1957. See DART. Marquardt –Cooper.

HAWK (Army/Marines)
Homing All the Way Killer. Surface-to-air interceptor missile. Length, 17 ft. Weight, over 500 lb. Solid propellant. Radar-homing guidance. Speed, Mach 2. Ceiling, low-level to 40,000 ft. SUPER HAWK under development. Initially operational in 1959. Used by Army, also Marine Corps and NATO allies. To be deployed in Europe, Far East, and Panama. Raytheon, prime; Raytheon, guidance; Aerojet-General, propulsion.

HAWKEYE
See SAINT.

HAYSTAQ (U.S. Patent Office/NBS)
Project for the development, test, and evaluation of an experimental search system which makes use of general-purpose computers.

HEDGEHOG
Antisubmarine rocket-launching device capable of firing 24 missiles. Used in World War II by U.S. Navy and British Royal Navy.

HEDGEHOG (Japan)
Japanese version of ship-to-submarine depth charge used by British Royal Navy in World War II. Rocket propelled. Range, 3 miles.

HELIODYNE
Space vehicle proposed by Northrop to explore solar system. Solar powered. Capable of carrying 70-kg payload. Can be lofted into any desired orbit to explore all points of the solar system. Described at Eleventh International Astronautical Congress, Stockholm, Sweden, 1960.

HELIOS II (NASA)
Also HELIO. A chemonuclear booster capable of lofting 230,000-lb payload into a 300-mile orbit or of soft-landing 60,000 lb on

the moon. Cost estimated at $1.5 billion. In R&D. Proposed by G.D.-Astronautics in 1960.

HELLER (Canada)
Antitank rocket with an unusual propulsion principle. Has been tested by the U.S. Army.

HELMET (ARPA)
An anti-ICBM program in government laboratories. In earliest R&D. Aim: development of a system to destroy enemy ballistic missiles in their terminal phase.

HERMES (Army)
Surface-to-surface military research vehicle. Liquid propellant. Range, 90 nm. Capable of pinpoint accuracy. Vehicle developed in 1950; canceled in June 1955. HERMES A-2 range, 75 miles. G.E.

HERO (Navy)
Study of BuWeps of electromagnetic radiation and its effect on storing, handling, and delivering missiles. In R & D.

HETS (Air Force/NASA)
Hyper Environmental Test System. A combination of two parallel programs by Air Force (SYSTEM 609A or BLUE SCOUT) and NASA (SCOUT). Low-cost vehicle for Air Force testing of ballistic missile equipment and NASA general-purpose tests in high-altitude environmental soundings.

H-F (NASA)
Designation for hydrogen and fluorine used as propellants, e.g., 15,000-lb-thrust liquid fluorine-liquid hydrogen engine. In breadboard stage. Bell.

HI-ALTITUDE SAMPLER (AEC)
Two versions: RT-2 which uses VIPERs 1 and 2; and HAS, which uses VIPERs 1 and 2 and a NIKE. Designed by Sandia.

HIBAL (AEC)
Balloon to take high-altitude air samplings in Australia.

HIGH CARD
See GENIE.

HIGH CUE (Air Force)
A continuous study of cloud formation and convective currents
in the atmosphere carried on at Flagstaff, Arizona. Aerophysics
Lab/Air Force/M.I.T.

HIGH POWER (DOD)
Microwave radar research program. Study of sensitive radars
for tracking ballistic missiles in the 1965-70 time period. Cornell Aeronautical Laboratory.

HILO
Ground-launched target. Ramjet engine. Greater speed than
current target drones. Recoverable. Suggested to both Air Force
and Army. Bendix/Republic/Marquardt.

HIPAR (Army)
HIgh Power Acquisition Radars for NIKE-HERCULES. G. E.

HIPERNAS I, II
Inertial guidance developed by Bell Aircraft Avionics.

HIROC (Air Force)
Single-stage research rocket. Post-World War II project. Vehicle had swiveling motors for directional control of the entire
missile, a separable nose cone, and lightweight structure. Last
MX-774 was launched in 1948. Was forerunner of ATLAS. Convair (now G. D. - Astronautics).

HI-STEP
High-speed Integrated Space Transportation Evaluation Program. A means of optimizing systems design by the use of an
IBM 7090 digital computer. Many candidate systems, and within
a selected system, many combinations of components, system
sizes, and operational procedures, can be evaluated and selected
and the system as a whole optimized using this method developed
by Lockheed, Burbank. Four subroutines are used: design, performance, cost, and optimization.

HIVOS
HIgh Vacuum Orbital Simulator. Test bed for satellites and
space vehicles. Lockheed Missiles and Space.

HOLY MOSES (Navy)
Air-to-surface missile. One of the HVAR family. A World War II vehicle, discontinued in 1955.

HONEST JOHN (Army)
Surface-to-surface, 762-mm artillery rocket. Military designation: M31; XM50. Length, 27 ft 3 in. Weight, 5900 lb. Solid propellant. Speed, Mach 1.5. Range, 12 miles. Thrust, 90,950 lb. Ballistic guidance; spin-stabilized. Engine: See M7A1. Model XM-50 is an improved HONEST JOHN made of aluminum with range of 15 to 30 miles and weight of 4900 lb. Length, about 25 ft. Douglas, prime for earlier M31; Hercules Powder, prime for M-50 version.

HOPI (Navy)
Air-to-surface missile with nuclear capability. Medium range. Designed for carrier-based aircraft. NOTS.

HOPI-DART
Research vehicle with capability of lofting 10-lb payload to a 40-mile altitude. Solid propellant.

HOPI-PLUS
Research vehicle with capability of lofting a 10-lb payload to a 60-mile altitude. Solid propellant. In R&D. Rocket Power/Talco.

HORIZON
A study conducted by the FAA at the request of President John F. Kennedy. Purpose: to make a statement of national aviation goals for the period to 1970. The results of this study were published in a report entitled "Report of the Task Force on National Aviation Goals: Project Horizon," FAA, September 1961. See also BEACON.

HORIZON (Army)
Proposed lunar base study.

HORIZON SCANNER I, II, III
Sensors for vertical reference data using visible radiation. Model I, used in satellites. Weight, 10 lb. In production. Model II, used in DISCOVERER. Weight, 10 lb. Operational. Model III, under development. Adv. Tech. Labs.

HOT POINT (Navy)
Antiaircraft missile. Nuclear warhead. In R&D.

HOUND DOG (Air Force)
Air-to-surface missile. Military designations: GAM-77; WS-131B. Length of vehicle, 42.5 ft. Launch weight, about 10,000 lb. Range, 500 miles. Speed, Mach 1.7. Air-breathing. Developed as an interim weapon before development and production of the SKYBOLT. Engine: See J-52. North American, prime; Autonetics, guidance; Pratt & Whitney, propulsion.

HSS-R 80/100 (Italy)
Surface-to-surface tactical missile. Solid propellant. Free guidance. Operational. Deployed with Egyptian troops. Boscarri/Meccanemica/Hispano-Suiza.

HTRE
Heat Transfer Reactor Experiments. Series of experiments to determine feasibility of direct-air cycle system. Conducted by G.E. from 1956 to 1961.

HTV (Air Force)
Hypersonic Test Vehicle. Used to test parachutes and drag devices at supersonic speeds, also to loft experimental payloads into the upper atmosphere. Designed and developed by Aerophysics Development, ARDC cognizance. Two- to six-stage configurations. Two-stage configurations: Length, 12.9 ft. Weight, loaded, 231 lb. Payload, about 10 lb. Speed, Mach 7 to Mach 10. Solid propellant. Three-stage configuration: Made of an HONEST JOHN, a T-40, and a NIKE. Research in heat transfer, stability, and drag at hypersonic speeds performed with aid of special nose cone. Speed, Mach 9 at 15-mile altitude. Four-stage configuration: Length, 35.7 ft. Weight, 2800 lb. Speed, Mach 10.4. Composed of HONEST JOHN or two NIKEs as first stage; a NIKE second stage; a Thiokol T-40 third stage; and a Thiokol T-55 fourth stage. Five-stage configuration: Developed by Pilotless Aircraft Research Station of NASA to obtain spacecraft and missile information at speeds of approximately Mach 16. Six-stage configuration: Reaches speeds of Mach 22. Composed of HONEST JOHN, first stage; and fourth, fifth, and sixth stages (Thiokol T-40, T-55, and a 5-in. spherical rocket motor) in a single housing.

HTV-1, 2 (Air Force)
Hypersonic test vehicle developed by Curtiss-Wright to test DYNA-SOAR scale models.

HUGO (ONR/Weather Bureau)
Instrumentation payload lofted to altitude approximately 100 miles in a long-range project to improve weather prediction. Payload weight, 38 lb. NIKE-CAJUN booster. Camera capable of photographing cloud cover over an area of 500,000 square miles.

HUMBRO (USCONARC)
Project which coordinates studies being made by the Human Resources Research Office. George Washington University.

HUMMINGBIRD (Army)
Advanced VTOL high-altitude airplane. Capable of hovering and of moving forward at jet speed. Tests jet ejection lift principle. Lockheed-Georgia.

HURRICANE HUNTER
See TIROS III.

HUSTLER (Air Force)
Also called BELL HUSTLER. Engine for upper stages of THOR-AGENA and ATLAS-AGENA boosters.

HVAR (Navy)
High Velocity Aircraft Rocket. Originally HOLY MOSES. Early air-to-air rocket used in World War II. Production stopped in 1955.

HYDRA (Navy/NOTS)
A program initiated by the Navy (NOTS) to determine the feasibility of launching from the surface of the water. Among advantages of ocean launching are mobility, freedom of site selection, simplicity of handling and logistics, and absence of political problems. The ocean launch pad is free, water-cooled, and self-healing. For nuclear power plants, radiation hazards are reduced. Solid propellants were used. Navy considers the HYDRA concept of possible use for future space launchings.

HYDRA I
Ocean-launched, solid-propellant vehicle developed by Navy. Successfully fired 18 March 1960 at Point Mugu to demonstrate the feasibility of launching from the ocean's surface. HYDRA I was lofted 60 ft using a 2.25 AEROSCAR rocket. Weight, 150 lb. Length, 70 in. Diameter, 10 in. (max). Width, 26 in. including fins. Thrust, 890 lb for 0.45 sec. Underwater ignition.

HYDRA II
> Ocean-launch experiment. Length, 40 ft. Weight, 10 tons. Tested successfully near San Clemente Island, Calif., in April 1961. Plans call for use of SCOUT rockets to conduct full-scale tests. Six SCOUT rockets have been ordered (March 1961).

HYDRA II (Navy)
> Series of 18 underwater chemical high-explosive detonation tests to determine nuclear antisubmarine weapon effects. Unrelated to HYDRA sealaunch concept. Conducted in July 1961 by Radiological Defense Laboratory.

HYDRA VIPER (Navy)
> A more advanced HYDRA vehicle launched from the surface of the water. Capable of propelling scientific payloads into the ionosphere. Consists of an ALGOL SCOUT booster and a solid-propellant second stage. See also HYDRA.

HYLAS-STAR
> Vehicle capable of boosting over 12,000-lb payload to escape velocity. Designed as a third stage for use with TITAN II. Aerojet-General.

I

ICBM
> InterContinental Ballistic Missile.

ICCM
> InterContinental Cruise Missile.

ICEF
> International Cooperative Emulsion Flight. An over-all study of cosmic rays by about 25 universities and research institutions from every continent except Antarctica. Sponsored by National Science Foundation and Office of Naval Research.

ICEWAY (Air Force)
> Research program to explore the possibility of using thick ocean ice in polar regions as airfields.

ICONORAMA (Air Force/NORAD)
> Data-processing and display system which records the movements of multiple airborne objects and projects their paths on a graphic display. Ling-Temco Electronics.

IFF

Identification Friend or Foe. Term used by Army.

IFR

Instrument Flight Rules. Term used for operating (especially landing) an aircraft safely by instruments.

ILAS (U.S. Patent Office)

A computer using row-by-row coding. Makes use of random interfix.

IM

Interceptor Missile.

IM-70

See TALOS.

IM-99A, B

See BOMARC.

IMPACT

Integrated Management Planning And Control Technique. Lockheed management program applied to POLARIS.

INFORMER

A general-purpose computer designed for all-weather, all-terrain use by Army. Fits into a $2\frac{1}{2}$-ton truck. Also may be adapted to logistic problems, missile checkout, intelligence data storage and retrieval, map compilation, and damage assessment. IBM.

INGSPANO

English-Spanish dictionary compiled by machine translation. University of Mexico.

INJUN (Navy)

Solar-powered radiation satellite to be second piggyback on TRANSIT IV. See also GREB, TRANSIT. James Van Allen, University of Iowa.

INSSCC (ARPA)

Interim National Space Surveillance Control Center. See SHEP-HERD.

INTERDICT
INTERference Detection and Interdiction Countermeasures Team. An operation charged with silencing r-f interference at missile sites. Capehart.

INTERNATIONAL I
See U.K. I.

INTRUDER (Navy)
Navy project. Contract to Grumman.

IRASER
Optical devices capable of picking up light waves in the infrared region, amplifying their effects, and thus determining the nature of remote objects. See also MASER, LASER, QUASER. Considerable importance is attached to these new concepts in instrumentation since they might well supersede radar. Hughes, also many other companies.

IRIS (NASA)
Single-stage, solid-fueled sounding rocket. Capable of lofting 100-lb payload to 190-mile altitude. Operational.

IRIS (Navy)
Sounding rocket. Solid-fueled. Capable of lofting a 100-lb payload to a 200-mile altitude. Similar to AEROBEE-HI. New version to have three fins to make possible launching of the rocket from a number of new launch sites. Original rocket had four fins. Atlantic Research.

ISKUSTVENYI SPUTNIK ZEMLI (U.S.S.R.)
Artificial Earth Satellite. Full Russian designation for SPUTNIK.

ISTAR
Image Storage Translation And Reproduction. Space reconnaissance system small enough to be carried by SCOUT-class rocket and capable of taking detailed, high-resolution photographs, storing the images, and transmitting them to earth on command. Has applications in earth's atmosphere as well as in space. Not a TV or similar system. In R&D. Ling-Temco Vought.

ITASPANO
Italian-Spanish dictionary compiled by machine translation. University of Mexico.

J

J-I (U.S.S.R. - Air Force/Army/Navy)
Surface-to-surface target drone. Solid propellant. Length, 26.85 ft. Weight, 9700 lb. Radio guidance. 2650-lb high-explosive warhead. Range, 375 miles at Mach 0.65. Ceiling, 21,300 ft. Operational.

J-2 (NASA)
A single-chamber engine. Liquid-hydrogen—liquid-oxygen propellant. Thrust, 200,000 lb. Under development by Rocketdyne. Will be clustered (four or six) for the SATURN C-3 second stage.

J-2 (U.S.S.R. - Air Force/Army/Navy)
Surface-to-surface target drone. Length, 36.7 ft. Weight, 16,100 lb. Solid-propellant boosters. 2200-lb nuclear warhead. Radio guidance. Range, 520 miles at Mach 0.87. Fired from submarines. Fitted with cargo nose container to carry supplies (3200) to troop units.

J-3 (U.S.S.R. - Navy/Army)
Surface-to-surface IRBM missile. Solid booster, ramjet sustainer. Radio guidance. Nuclear capability. Length, 37.2 ft. Weight, 20,900 lb. Range, 1450 miles at Mach 1.15. Operational with the Soviet army and at least seven Baltic fleet cruisers.

J-44
Turbojet engine; 1000-lb thrust. Fairchild.

J-52
Rocket engine (jet) developed by Pratt & Whitney.

JAGUAR (NASA/Air Force)
Test program to explore capability of air-launched sounding rocket operations in remote areas. Three- and four-stage rockets launched from B-57 bombers to loft instrumentation packages to over 500-mile altitude. In R & D. ARDC project control.

JAKTROBOT (Sweden)
See ROBOT 304.

JANET (Air Force)
Research vehicle. Part of FIREFLY project. Operational.

JASON
  See ARGO E-5.

JAVELIN
  See ARGO D-4.

JB
  Jet Bomb.

JB-2
  See LOON.

JEIPAC
  Japan Electronic Information Processing Automatic Computer.
  Japanese data-processing system. See also JICST.

JERICHO LUTIN (France)
  Surface-to-surface missile. Length, 4.25 ft. Diameter, 16 in.
  Weight, 33 lb. Ramjet engine. Range, 3 miles. Speed, 280 mph.
  BTZ.

JEZEBEL (Navy)
  Buoys used with Navy airborne detection system, JULIE. Locate
  submarines by triangulation. Lockheed-California.

JICST
  Japan Information Center of Science and Technology.

JINDIVIK (Australia)
  Target drone used by Australia, Great Britain, and Sweden.
  Length, 23.3 ft. Weight, loaded, 2900 lb. Payload weight, approximately 380 lb. Altitude, 10 miles. Speed, about 575 mph. Liquid
  propellant. Designed by Government Aircraft Factories (Australia) to specifications by the British Ministry of Supply.

JOBTRAIN (U.S. Army-Signal Corps)
  Electronics training program.

JOSHUA (Air Force)
  Classified Air Force project to develop lightweight ICBM
  powered with liquid-propellant rocket engines. Marquardt has
  $100,000 contract. Edwards Rocket Test Agency working on the
  development. See also MIDGETMAN, SHRIMP, SICBM.

JOURNEYMAN
  See ARGO D-8.

JPL (NASA)

Jet Propulsion Laboratory, Pasadena, Calif. Major task is development of lunar and interplanetary probes such as RANGER, PROSPECTOR, SURVEYOR, and VOYAGER; however, not the manned lunar mission APOLLO.

JUDI-DART

Research rocket. Solid propellant. In use by several agencies. Rocket Power/Talco.

JULIE (Navy)

Submarine detection system used in conjunction with JEZEBEL. Consists of floating buoys using explosives for echo ranging. Airborne. Advanced development by Daystrom and Bell Telephone Laboratories.

JUNO II

Early deep-space booster developed by the Army (ABMA), now under NASA. First stage is a modified JUPITER; second stage, 11 scaled-down SERGEANTs; third stage, three scaled-down SERGEANTs; fourth stage, one scaled-down SERGEANTs. Capable of launching about 95-lb payload into 600-mile orbit. Used to launch PIONEER III and IV and EXPLORER VI, VII, and VIII. Phased out at end of 1961. Marshall Center/Chrysler, prime; Ford Instrument, guidance; Rocketdyne/JPL, propulsion.

JUPITER (Army)

Surface-to-surface IRBM rocket. Liquid propellant. Length, 60 ft. Weight, 90,000 lb. Inertial ballistic guidance. Speed, over 10,000 mph. Range, more than 1500 miles. Thrust, 150,000 lb. Operational. Deployed in Italy. Considered for use in Turkey. In competition with POLARIS for more general use by NATO. Engines: See XM83-4; XM148. ABMA/Chrysler, prime; Ford Instrument, guidance; Rocketdyne, propulsion.

JUPITER C (Army)

Satellite-launching vehicle for many of the EXPLORER satellites. Length, 69 ft 2 in. Weight, 65,000 lb. Inertial guidance. Liquid-solid propellant. Speed, 19,000 mph. Range, 3500 miles. Thrust, 78,000 to 83,000 lb. Only related to JUPITER for funding purposes. ABMA/Chrysler.

K

KAN-1,2

See LITTLE MOE.

KAPPA (Japan)
Rocket project performed by Science Council of Japan for the International Geophysical Year. Designed specifically for upper-atmosphere sounding. Initially, purpose was to develop a multi-staged, solid-propellant, instrumented, ground-launched sounding rocket capable of climbing 80 miles with a 13-lb payload. Several models were developed. See KAPPAs III through X. See also SIGMA.

KAPPA III (Japan)
Single-stage sounding rocket. Length, 107.5 in. Diameter, 5.1 in. Weight, 100 lb. Maximum altitude, 20 miles. Payload, about 18 lb.

KAPPA V (Japan)
Single-stage sounding rocket. Diameter, 4.9 in. Weight, 95 lb. Payload, about 13 lb.

KAPPA VI (Japan)
Two-stage sounding rocket. Length, 19.3 ft. Weight, 595 lb. Maximum altitude, 37.2 miles. First launched October 1958. Nose cone with solar spectroscope recovered from water after 6 days. Several launchings during the International Geophysical Year. Purpose: to measure solar radiation, ion density, cosmic rays, air pressure, and temperature. Cost per copy, about $46,000, excluding R&D. Yugoslavia bought six for spring 1962 delivery. Fuji Precision Machinery Co. and Mitsui Bussan Co.

KAPPA VII (Japan)
Single-stage sounding rocket. Uses K-420 solid booster. Thrust, 20,000 lb. Length, 6 m. Diameter, 428 mm. Some 70 test firings made, starting July 1957.

KAPPA VIII (Japan)
Two-stage sounding rocket. Length, 33 ft. Diameter, 428 mm. Weight, 3256 lb. Altitude, 100 to 125 miles. Thrust, 20,000 lb. Payload, about 77 lb. Five firings from March to September 1960. Used for studying cosmic rays. Japanese FY 1961 budget, $465,000 for KAPPAS VII and IX.

KAPPA IX (Japan)
Three-stage booster. Length, 40 ft. Weight, 3300 lb. Capable of lofting 13.2-lb payload over 200-mile altitude. Launch scheduled for 1961. Possible satellite launcher.

KAPPA X (Japan)
Booster scheduled for launch in 1961-62. Thrust, 22,000 lb. Capable of lofting 44- to 66-lb payload to over 200-mile altitude. Cost, $2 million. Will carry TV camera, radar, and transmission instrumentation.

KATHY (Air Force)
Research rocket capable of lofting 8- to 35-lb payload. Part of FIREFLY project. Operational.

KATIE (Air Force)
Surface-to-surface and antisubmarine torpedo. Launched by 16-in. guns. Nuclear capability. In R&D.

KATYBIRD KDB-1 (Navy/Army)
Target drone built by Beech Aircraft. Length, 15 ft. Launch weight, 593 lb.

KATYBIRD KD2B-1 (Navy/Air Force)
Target drone used for training. Military designation, Q-12. Capable of flight from 5- to 70,000-ft altitude. Guidance system permits target to fly to predetermined altitudes and speeds.

KCAT-25 (U.S.S.R. - Army)
Antitank missile, believed operational.

KDA-1, KDA-4 (Navy)
Military designation for aerial target drone FIREBEE.

KDB-1
Also known as CARDINAL. See KATYBIRD KDB-1.

KD2B-1
See KATYBIRD KD2B-1

KD2R-5 (Navy)
Standard target drone used by Navy. Same as OQ-19 and OQ-19D,E used by Army and OQ-19B, used by Air Force. Radioplane.

KD2U-1 (Air Force)
Target based on REGULUS II supersonic, aerodynamic, surface-to-surface missile. Now the target for the BOMARC B at Eglin AFB. The solid-propellant booster used for the drone is the M-34, formerly called ZEL and MEGABOOM.

KD6G-2
Target drone developed for the Navy.

KDM-1
See PLOVER.

KINGFISHER (Army)
Target drone based on X-7. Military designations: Q-5, XQ-5.
Lockheed Missiles and Space; Marquardt, propulsion (ramjet);
Thiokol, propulsion (solid).

KISHA
Solid motor. Length, 72.1 in. Diameter, 6.5 in. Weight, loaded
145 lb; burnt 42 lb. Thrust, a nominal 48 lb for 4.8 sec. Rocket
Power.

KISHA-JUDI
Two-stage sounding rocket. Weight, 221 lb. Maximum altitude,
480,000 ft, with highest second-stage velocity, 3271 ft/sec.
Rocket Power.

KIVA-DART
Research rocket. Payload, 25 to 55 lb. Solid propellant. Operational. Rocket Power/Talco.

KIWI A, A', and A3 (AEC)
Experimental nuclear reactors for use in nuclear rocket program. Tested in Nevada in 1960 and 1961. Reactor uses gaseous
hydrogen as propellant and water as a coolant. NERVA engine
for use in ROVER program to be based on these experiments.

KIWI B, B1, B1A (AEC)
Experimental nuclear reactor for use in development of NERVA
(ROVER program). To use liquid hydrogen as propellant and
liquid hydrogen as a coolant. Flight tests planned for 1966-67.
Flight-test program called RIFT. Resulting engine to be candi-
date for future upper stages of SATURN and NOVA.

KOMET D (U.S.S.R. - Air Force)
Air-to-surface turbojet-propelled missile. Radio-radar guid-
ance. Length, 15.0 ft. Weight not known. Solid propellant.
Thermo-nuclear warhead. Range, 93 miles at Mach 0.85.
Launched at 50,000-ft altitude from BEAR or BISON bomber.

KOMET I (U.S.S.R./Soviet Missile Command)
Surface-to-surface missile. Solid propellant. High-explosive and nuclear warhead. Range, 100 miles. Speed, 3000 mph. Reported to be operational with U.S.S.R. Army. Scheduled to put KOMET I aboard submarines and surface ships. May also be in operation with Red China Navy.

KOMET II (U.S.S.R./Soviet Missile Command)
Surface-to-surface IRBM. Solid propellant. Length, 42.4 ft. Weight, 41,350 lb. 550-lb high-explosive warhead. Inertial guidance. Range, 620 miles at Mach 7.5. In production in U.S.S.R. and East Germany. Reports claim about 3200 KOMET I and II produced each month. Operational with Army and from surface vessels in calm seas. In R&D for submarines.

KOSMICHESKIJ KORABL-SPUTNIK I (U.S.S.R.)
KORABL-SPUTNIK designation given to animal-carrying satellites only. See SPUTNIK IV.

KOSMICHESKIJ KORABL-SPUTNIK II (U.S.S.R.)
See SPUTNIK V.

KOSMICHESKIJ KORABL-SPUTNIK III (U.S.S.R.)
See SPUTNIK VI.

KOSMICHESKIJ KORABL-SPUTNIK IV
See SPUTNIK IX.

KOSMICHESKIJ KORABL-SPUTNIK V
See SPUTNIK X.

KOSMICHESKIJ KORABL-SPUTNIK VI
See SPUTNIK XI.

KOSMICHESKIJ KORABL-SPUTNIK VII
See SPUTNIK XII.

KWIC
Key Word In Context. An indexing system using a computer, e.g., IBM 704, and other data-processing equipment. Makes list of publications by all key words in their titles. Thus, a "quick" indexing system.

L

L-1

Liquid-propellant rocket engine to follow the F-1. Will have 10 to 12 million-lb-thrust range.

LABOR OF LOVE

Liquid-fueled rocket built by Marshall Kriesel, a student from Owatonna, Minn., from 1956 to 1961. Length, 11 ft. Fired successfully at China Lake, Calif., in 1961.

LABS (Navy)

Low Altitude Bombing System. Used by Navy for loft bombing.

LACE (Air Force)

Liquid Air Cycle Engine. Power plant concept under study for proposed Air Force Aerospace Plane. Liquid hydrogen is used to cool and liquefy air as the plane flies through the atmosphere. The liquid air is then pumped to the combustion chamber and burned with the liquid hydrogen which was used to liquefy it. See also NULACE. Marquardt.

LACROSSE (Army)

Surface-to-surface field artillery missile. Length, 19.2 ft. Weight, 2300 lb. Range, 20 miles. Speed, Mach 0.9. Radio-command guidance. Solid propellant. Operational. Marine version canceled. Advanced LACROSSE R&D dropped. Engines: See XM10. Martin, prime, guidance; Thiokol, propulsion.

LADD

Low-Angle Drogued Delivery. A technique of delivery of nuclear weapons from low-flying aircraft. Replaces the earlier LABS delivery. After making a low-level run-in to the target, the pilot pulls up only slightly at bomb release to a peak altitude of several hundred feet. Then he half-rolls, decreases altitude, half-rolls again on the descent, pulls out, and continues at low level.

LAMP (Army)

Lunar Analysis and Mapping Program. Directed by Corps of Engineers. Phase I of the program calls for a 1 : 5 × 106 scale lunar topographic map based on available lunar photography. In Phase II, new information will be obtained by photography from balloons and by radio interrogation. Phases III and IV call for the design and development of a lunar orbiter and payload. It is expected that Phases III and IV will be merged into SURVEYOR.

LAMPRE I
Plasma-propulsion space engine developed by RCA. Engine uses UHF radio waves to accelerate charged particles.

LANCE
See WHITE LANCE.

LANGLEY RESEARCH CENTER (NASA)
Located at Langley, Va. Primarily concerned with space flight, advanced aircraft, and re-entry.

LARK (Navy)
Surface-to-air subsonic missile. Program was begun February 1945. Purpose: to develop an early training vehicle for launching crews. Discontinued in December 1950. Fairchild.

LARVA
Low-Altitude Research Vehicle. A ramjet research vehicle capable of Mach 2 to 3 for 10 min. Marquardt.

LARYNX (Great Britain)
Surface-to-surface guided weapon. Small airplane capable of carrying a 250-lb load of bombs at 200 mph for about 110-mile distance. Radio-controlled. Work began in 1927. Developed by British Royal Aircraft Establishment.

LASER
Light Amplification by Stimulated Emission of Radiation. Union of optics and electronics to allow carrying electromagnetic energy in the visible and IR regions. Planned for use in advance communications. See MASER, IRASER, QUASER. Hughes.

LAST DITCH (DOD)
Project to defend against missiles which might penetrate a NIKE-ZEUS defense system. Some damage in the target area accepted by the concept. Douglas has study contract.

LAURA (Sweden)
Antisubmarine rocket. Longer range than Bofors AURORA. Range, 450 to 1650 m. Cannot be fired from AURORA launcher. Bofors.

LAW (Army)
Light Antitank Weapon. Carried and fired by one man. Range, 10 to 500 yd. Solid propellant. In R&D. Flightex Fabrics—Hesse-Eastern Div., prime.

LAZY DOG (Air Force)
Air-to-surface antipersonnel missile. Length, only a few inches.

LENTICULAR ROCKET (Air Force)
A disc-cross section, "flying-saucer" configuration, aero-dynamic vehicle. May be a study of the aeroballistic missile concept since such a rocket could change its course to avoid antimissile missiles. Feasibility study contract for about $300,000 awarded to G. D.-Astronautics in 1960.

LEST
Management planning and control system devised by IBM. See also PERT, PEP.

LEWIS RESEARCH CENTER (NASA)
Located at Cleveland, Ohio. Primarily responsible for research in aerospace propulsion and power generation.

LEXINGTON
Nuclear-powered aircraft project started in early 1950's. Amount spent to date on nuclear-powered aircraft totals about $1 billion. LEXINGTON project being phased out.

LIBERTY BELL 7
Name given MERCURY capsule for the second U.S. astronaut's (Capt. Virgil Grissom's) flight in a ballistic trajectory, 21 July 1961. Capsule was modified version of FREEDOM 7, including a large observation window and a more advanced type of manual control in the rate command attitude control system.

LIGHTNING (Navy)
Research into the progressing state-of-the-art in high-speed data-processing systems, with development of a computer capable of operating at a speed of 10^{-9} sec or greater. RCA/Univac, prime.

LIL
Lunar International Laboratory. Proposed by Dr. Theodore von Karman at the IAF meeting in Washington, October 1961. Predicted LIL will be established on the moon within 25 years.

LINAFAC
Commercial automatic checkout and data-processing system. Lockheed Missiles and Space.

LINS

Lightweight Inertial Navigation System. Weighs 124 lb and occupies less than 3 ft^3 of space. Can be applied to both short- and long-range military requirements and to most of the commercial operations of extended range and flight times. Also applicable to various classes of missiles. Nortronics.

LITTLE JOE (NASA/Army)

Solid-propellant booster used to test MERCURY escape capsule, capabilities, design concepts, and recovery system. North American prime; Langley; Thiokol.

LITTLE JOHN

Surface-to-surface, unguided heavy field artillery rocket (318 mm). Military designations: XM47; XM51; M-51. Length, 12 ft. Launch weight, 780 lb. Ballistic guidance; spin-stabilized. Solid propellant. Speed, supersonic. Range, 10 to 13 miles. Supplement to HONEST JOHN. Operational, 1961. Marine version canceled. Engine: See X-235. Emerson Electric, prime; Hercules Powder, propulsion.

LITTLE MOE (Air Force)

Surface-to-air guided missile. Military designations: KAN-1; KAN-2. Produced by McDonnell in 1944.

LN· I, II, III, IV (NASA)

Four stages proposed for NOVA vehicles. LN-I is composed of eight Rocketdyne F-1 engines. LN-II: two Rocketdyne F-1 engines. LN-III: four Rocketdyne J2 engines. LN-IV: six Pratt & Whitney RL-10 engines. Proposed fifth stage: the APOLLO command center. See also NOVA.

LOBBER (Army)

Short-range vehicle with range about 25 miles or under. Parallel project to ADAM (Army). Can deliver rations, ammunition, and medicines to front-line troops when and where necessary. Three-man team in field can hand-carry LOBBER. Also adaptable to offensive weaponry. Under study by G. D.-Astronautics.

LOBOS (NASA)

A booster study.

LOD (NASA)

Launch Operations Directorate. Responsible for coordination of all of the agency's launch operations. Also has assumed ABMA's obligated program to the Ordnance Missile Command for launching JUPITER, PERSHING, and REDSTONE missiles.

LOFTER
    See ARPA 162-61.

LOFTI (Navy)
    LOw Frequency TransIonospheric piggyback satellite lofted 21
    February 1961 with TRANSIT IIIB. Purpose: to test feasibility
    of a satellite relay for communication with submerged sub-
    marines by VLF radio. Re-entered atmosphere after operating
    for 6 weeks.

LOGBALNET
    Former communications system used for ballistic missile
    logistics. See COMLOGNET.

LOKI (Army)
    Antitank unguided rocket fired from helicopters. Solid propel-
    lant. Version of the German TAIFUN unguided barrage rocket.
    Bendix; East-Coast Aeronautics.

LOKI-DART (Navy)
    Solid-propellant research rocket capable of lofting 8-lb payload
    to 40-mile altitude. Navy to fire from shipboard. Marquardt-
    Cooper.

LOKI-WASP (Army/Navy)
    Solid-propellant research rocket capable of lofting 6-lb payload
    to 35-mile altitude. Operational.

LOMAR (Army)
    LOgistics, MAintenance, and Repair of satellites and space-
    craft. Proposed study. Canceled.

LONG ARM (WADD - Comm. Lab)
    Air-to-ground-to-air, long-range HF digital communication
    system. Hughes.

LONGLEGS
    Payload for ALOUETTE, a Canadian satellite for probing upper
    ionosphere. Joint Canada-NASA launching scheduled for 1962.

LONGSIGHT (ARPA/Air Force/Army)
    Unorthodox, creative studies and system analyses in missiles
    and spacecraft designed to discover projects needed for future
    military requirements. SPAD and PUT PUT are outgrowths of
    LONGSIGHT. Part of DEFENDER. R&D study contract to Lock-
    heed-Georgia.

LONG TOM (Australia)
Sounding rocket fired successfully several times. Awaiting government approval to launch small satellites (mid-1961).

LOOKOUT (Canadian Armament R&D Establishment)
Canadian project to detect, track, and record missile flights. Classified. CARDE.

LOON (Navy)
U.S. version of the German air-breathing missile, V-1. Military designation JB-2. Built immediately following World War II, as a missile to be launched from submarines. Discontinued in September 1952.

LORRAINE (ARPA)
Basic research in advanced energy-conversion techniques and energy storage and collection.

LOTS (Army)
Logistic Over The Shore. 15-ton surface effect machine. Dimensions: about 35 ft by 62 ft. Purpose: to carry men and supplies over cleared roadways. See also GEM.

LOW BLOW (Navy)
Surface-to-air IRBM missile in early research. Ramjet propulsion.

LOW CARD (Air Force)
Evaluation of infrared detectors of missile launchings. See SMOKEY JOE.

LR44-RM-2(TD-174)
Liquid rocket motor used with SPARROW 3. Thiokol.

LR 87-AJ-3
First-stage engine used with TITAN I. Liquid propellant. Thrust, 300,000 lb. Aerojet.

LR 87-AJ-5
First-stage engine used with TITAN II. Liquid propellant. Thrust, 400,000 lb. Aerojet.

LR 89 NA-5 MA-2
Liquid-propellant booster used for ATLAS. Thrust, 360,000 lb. Rocketdyne.

LR 89 NA-5 MA-3
Liquid-propellant booster for ATLAS. Thrust, 389,000 lb. Rocketdyne.

LR 91-AJ-3
Second-stage engine used with TITAN II. Liquid propellant. Thrust, 70,000 lb. Aerojet.

LR 91-AJ-5
Second-stage engine used with TITAN II. Liquid propellant. Thrust, 70,000 plus lb. Aerojet.

LR-105 NA-5 MA-2
Liquid sustainer engine for ATLAS. Thrust, 360,000 lb. Rocketdyne.

LR-105 NA-5 MA-3
A 389,000-lb solid sustainer, liquid propellant, for ATLAS. Rocketdyne.

LR-119 (Air Force)
SATURN S-IV stage engine. Thrust, 17,500 lb. Pratt & Whitney.

LTV
Launching Test Vehicle.

LULU (Navy)
Air-to-surface or ASROC-launched nuclear depth charge. Operational. General Mills/Naval Ordnance Lab, prime.

LULUBELLE (Navy)
Surface-to-surface missile. Nuclear warhead. In R&D.

LUNA
Projected future booster whose size is approximately that of the Eiffel Tower. Thrust to be 10 million lb.

LUNIK I (U.S.S.R.)
Also called MECHTA. Moon probe launched 2 January 1959. Weight, 3245 lb. Payload weight, 796.5 lb. Carried scientific research instruments. Passed within 5000 miles of the moon to enter solar orbit. Aphelion, 123.1 million miles; perihelion, 91.2 million miles. In 450-day solar orbit.

LUNIK II (U.S.S.R.)
>   Moon probe launched 12 September 1959. Payload, 796.5 lb. Flew 236,876 miles to impact on moon. Purpose: to make radiation and magnetic measurements.

LUNIK III (U.S.S.R.)
>   Moon probe launched 4 October 1959. Payload weight, 613 lb. Probe orbited moon and photographed its unseen side. Apogee, 298,000 miles; perigee, 30,000 miles.

M

M-1 (U.S.S.R. -Army)
>   Surface-to-air missile. Length, 15 ft. Weight, 3300 lb. Three solid-propellant boosters. Liquid engine sustainer. HE warhead Inertial guidance. Velocity, Mach 1+. A training missile developed about 1947, now phased out. Alos used as a surface-to-surface tactical-type missile.

M-2
>   See CORPORAL.

M-2 (U.S.S.R. - Army/Navy)
>   Solid-fueled, antiaircraft missile similar to U.S. NIKE-AJAX. Length, 36 ft. Weight, without booster, 1500 lb. 26.5-lb HE warhead. Range, 35 miles at Mach-2 speed. Ceiling, 54,000 ft. Guidance IR, or radar, or both. Operational with ground defense forces and aboard Baltic cruisers. First seen in Moscow in 1957. New version reported to reach 150,000 ft. M-2 given credit for bringing down the U-2 reconnaissance plane.

M4E2
>   See LACROSSE.

M-5
>   See PERSHING.

M7A1
>   Steering engine, solid propellant, used with HONEST JOHN. Thiokol.

M16E1
>   Solid-propellant booster used with the MATADOR and MACE. Thiokol.

M30; M31-A1C
   Also XM50. See HONEST JOHN.

M-34 (Air Force)
   Solid-propellant booster for the KD2U-1 drone (formerly Navy
   REGULUS II). Rocketdyne. See MEGABOOM, ZEL, REGULUS II,
   KD2U-1.

M-55 (Army)
   Also called BOLT. Military designations: T-238; WS-T238. Sur-
   face-to-surface tactical missile. Diameter, 4 in. Solid propellant.
   Short range. Free-flight guidance. Chemical-agents warhead.
   Initially operational in 1960. Deployed with U.S. combat troops
   in the U.S. and overseas. Norris Thermador, prime.

M58E1,2 (Air Force)
   Military designations, TX-58-1,2. Sustainer engines for FAL-
   CON. Solid. Thiokol.

M-100 (U.S.S.R. - Air Force)
   Air-to-air unguided missile. Length, 13 ft. Weight, 990 lb. Solid
   propellant. 336-lb HE warhead. Range, 5 miles at Mach 0.96.
   Developed by German technicians in 1947. Used in the Arctic
   regions by the U.S.S.R. Anti-Air Defense Force (PVO).

M-100A (U.S.S.R. - Air Force)
   Air-to-air missile. Solid propellant. IR guidance. Range, 3.5
   miles. Speed, Mach 2.5 Control by jet deflection. Also in an
   unguided version. Similar to U.S. FALCON (GAR-3).

M-101
   See T-1 (U.S.S.R.).

M-102
   See T-4 (U.S.S.R.).

M-103
   See T-2 (U.S.S.R.).

M-104
   See T-3 Mk 1 (U.S.S.R.).

M-109 (Army)
   Missile developed by Ryan Aeronautical.

M-113 (Army)
Missile launcher for MAULER, developed by Food Machinery.

MA-1
See GENIE.

MA-1 to -14 (NASA)
Capsules for the MERCURY Man-in-Space program. ATLAS booster. MA-5 carried chimpanzee Enos on orbital flight 29 November 1961. Capsule brought down after two orbits because of overheating of two inverters.

MACE (Air Force)
Turbojet-powered, surface-to-surface tactical missile. Military designations: TM-76A; TM-76B; WS-309A. Air-breathing. Length, 44 ft. Weight, 13,000 lb. Solid propellant. Speed, over 650 mph. Range, 650 nm (A model); 1200 nm (B model). MACE A deployed with U.S. troops in West Germany; MACE B to be deployed on Okinawa. Martin, prime; Goodyear/AC Spark Plug, guidance; Thiokol/Allison, propulsion.

MAD
Maintenance Assembly and Disassembly. Term used by the AEC for maintenance assembly and disassembly procedures or locations for ROVER and other programs.

MAD (Navy)
Magnetic Anomaly Detection. Location of distortion of the earth's magnetic field caused by passage of a submarine.

MADDAM
MAcro-module Digital Differential Analyzer Machine. Miniaturized computer made by Burroughs. Weight, 13 lb. Size, 3 × 6 × 11 in. 97 w. Serial, binary, digital differential analyzer.

MADRE (Navy)
Magnetic Drum Receiving Equipment. Radar system capable of vision over the horizon. Bounces low-frequency signals off the ionosphere, stores them, and then compares them on magnetic-drum receiving equipment. In R&D. NRL.

MALAFAC (France - Navy)
Surface-to-surface missile. Solid-liquid propellant. Length, 20.5 ft. Range, about 25 nm. Radio-command guidance. HE warhead. Several test missiles have been launched at sea. To be installed

aboard French surface warships. Société Industrielle d'Aviation
Latecoere.

MALAFON (France - Navy)
Surface-to-underwater homing torpedo. Considered similar to
U.S. Navy ASROC. Sonar and radio-command guidance. Solid
propellant. HE warhead. Range, more than 10 nm. Operational
in 1960. Deployed first aboard ASW Command Ship La Galisson-
ière. Société Industrielle d'Aviation Latecoere.

MALKARA (Australia - Army)
Two-stage surface-to-surface antitank missile. Solid propellant.
Launch weight, 205 lb. Wire-guided. Speed, 400 mph. HE warhead. Range, about two miles. Deployed with British combat
troops. Operational. Government Aircraft Factories.

MALLAR
Manned Lunar Landing And Return. Term used by Ling-Temco
Vought (formerly Chance-Vought) for proposed manned lunar and
return mission in an early 1961 proposal. Program estimated to
cost $3.1 billion.

MAN
Microwave Aerospace Navigation. Remote control system for
recovering space vehicles. Sperry-Phoenix.

MAN-HIGH (Navy/ONR)
Also called DAEDALUS. Project to conduct upper atmosphere
(above 80,000 ft) research from balloons. Telescopes and spec-
trometers used to measure infrared spectrum of planets,
especially Venus. Cmdr. Malcom Ross is chief researcher and
pilot. A continuous program. Johns Hopkins University.

MAN-HIGH I, II, III (Air Force - Aerospace Medical Center)
A series of manned balloon flights for upper atmosphere research, especially in space medicine and human factors. Chief
researcher and pilot was Lt. Col. David G. Simons. Maximum
altitude, over 100,000 ft for MANHIGH II and III; over 32-hr
continuous flight for MANHIGH II.

MAN-ON-MOON (NASA)
Term used for follow-on program to APOLLO. Lunar mission
with 30-day stay for crew in permanent moon base. NOVA is
possible booster.

MAPCHE
Mobile Automatic Programmed CHeckout Equipment. Contract for $878,000 awarded to RCA by G. D.-Astronautics (formerly Convair). To be used at the "F" series ATLAS sites.

MARCO POLO
Nickname for VIKING during its early development.

MARINER (NASA)
Series of seven planetary probes under development by Jet Propulsion Laboratory (prime). First two probes, designated MARINER A, to be flybys to Venus and Mars. These were reduced in weight, from the 1100-lb spacecraft originally planned, to 450-lb probes in order to be launched by ATLAS-AGENA B boosters at the optimum times, 16 August 1962, for Venus flyby, 13 November 1962, for Mars shot. CENTAUR booster first intended for MARINER series will not be available by these dates. MARINER A to consist of three 1100-lb spacecraft. Scheduled missions are one lunar flight and two Venus flights in 1964. MARINER B to be two similar spacecraft, to make flights to Mars in 1964. Studies by these probes to relate to atmospheric constituents, pressures and densities, structure, dynamic features; planetary ionosphere, if any; interaction of energetic particles with planet's atmosphere and fields; electric, magnetic, and gravitational fields; planet's structure and surface features; extraterrestrial life, if any; and conditions in the region of transfer orbit.

MARK 2
See MERCURY.

MARS
Manned Astronautical Research Station. Three-man space laboratory proposed by G. D.-Astronautics for testing life-support systems. Would orbit earth for periods of several weeks at 200-mile altitude.

MARS (Air Force)
Military Affiliated Radio System. Group of 15,000 amateur radio operators, both military and civilian. Supplement to regular communications between Air Force installations.

MARS (Air Force)
Mobile Atlantic Range Stations. Two ships to be outfitted for Atlantic Range service. Sperry-Gyroscope.

MARS (NASA)
Multi-Aperture Reluctance Switch. A record storage unit designed for the Orbiting Astronomical Observatory (OAO). Largest random access memory of its kind. Product of three years' R&D by IBM.

MARSHALL SPACE FLIGHT CENTER
See GEORGE C. MARSHALL SFC.

MARSHMALLOW
Study of use of certain materials or devices for absorbing radiation.

MARUCA (France - Navy)
Surface-to-air solid-liquid—fueled missile. Length, 15 ft. Range, about 10 nm. HE warhead. A development vehicle for MASURCA. Used with French ship, Ile d'Aleron. Operational, for training. Ruelle Naval Arsenal.

MARVELETTE (Army)
Prototype aircraft, Marvel STOL program. Mississippi State University.

MASALCA (France - Navy)
Surface-to-air, air-breathing missile, deployed on French cruisers. Launch weight, 6000 lb. HE warhead. Range, about 65 miles. Solid propellant. Beam-rider guidance. Operational in 1961. Société Industrielle d'Aviation Latecoere.

MASER (Air Force)
Microwave Amplification by Stimulated Emission of Radiation. Many laboratory studies in progress involving use of light amplification for detecting and tracking objects. MASER will perform functions of radar but for much greater ranges. One study contract to Columbia University. See LASER, IRASER, QUASER.

MASTIF (NASA)
Multi-Axis Space Test Inertia Facility. A simulator used by MERCURY pilots during their training, in which the astronaut is tumbled through three axes simultaneously. The astronaut also practices manual control through using compressed air reaction jets to stop the device, watches the panel display, and uses the communication system. MASTIF is one of five simulators of varying sorts designed especially for the MERCURY program.

MASURCA (France - Navy)
Surface-to-air missile. Solid propellant. Length, about 19 ft. Launch weight, 3200 lb. Range, more than 20 nm. Speed, over Mach 2. Radar beam-rider guidance. Operational in 1960 aboard helicopter carrier, Jeanne d'Arc. Also on new pocket cruisers and squadron escorts weighing over 3000 tons. Considered to be the top air defense weapon of French navy. Ruelle Naval Arsenal.

MATADOR (Air Force)
Turbojet surface-to-surface missile. Military designations: TM-61A; TM-61C; MX-771; XSSM-A-1. Length, over 39 ft. Weight, loaded, 12,000 lb. Payload weight, 3000 lb. Range, 650 miles. Speed, Mach 0.9. Air-breathing. Solid-propellant booster. Vehicle being turned over to West Germany; also assigned for service in Formosa. TM-61B now called MACE, which is several feet longer than TM-61A and TM-61C. Martin, prime; Thiokol/ Allison, propulsion.

MATRA (France)
Société Générale de Mécanique-Aviation-Traction.

MAULER (Army)
Antiaircraft and antimissile missile designed to intercept incoming tactical missiles of the HONEST JOHN – LITTLE JOHN class. Also has capability of intercepting both fixed-winged and rotary-winged aircraft. Solid propellant. Radar guidance. Highly mobile. Launched from tracked vehicle by small crew. In R&D phase. To replace CORPORAL. Under consideration by NATO. Cost of development, about $350 million. G.D.—Astronautics, prime; Raytheon, radar; Burroughs, computers; FMC, vehicle.

MAXSECOM
MAXimum SEcurity COMmunications. Infrared communications system which uses IR beams instead of conventional radiowaves to transmit speech. Range, only 3 miles at present, but can be extended to 20 miles. Can be held by one man who "shoots" words out of the transmitter to the IR receiver. Completely immune to interference, detection, or jamming.

MAW (Army)
Antitank missile. Under development.

MB3-I
Engine which was standard in the DM-18 THORS, and used in the first 12 DELTA boosters. Develops 150,000-lb thrust.

MB3-II
Engine with 170,000-lb thrust. To be used in second group of DELTA boosters ordered by NASA to launch TIROS, RELAY, SYNCOM, and TELESTAR.

MB-1, MB-17
See GENIE.

MCM (Air Force)
Missile Carrying Missile. Surface-to-air missile designed to carry several smaller target-seeking, antiaircraft, supersonic missiles. R&D.

MECHTA (U.S.S.R.)
Russian term meaning "dream." See LUNIK I.

MEGABOOM (Navy)
Solid-propellant booster, designed as the rocket for pushing the test sleds at Holloman AFB. Produced by Astrodyne. Another configuration of the MEGABOOM evolved into the REGULUS II booster system and was called ZEL. This configuration is now called M-34 and is produced by Rocketdyne.

ME (IGOR) (U.S.S.R. – Army)
Antitank missile. Length, 4 ft. Weight, over 10 lb. Solid propellant. Range, over 2 miles at Mach 0.7. No guidance. 3.75-lb HE warhead. Similar to U.S. BAZOOKA. In production since 1946.

MELVA
Military Electronic Light VAlve. A projection system being developed for use with MIDAS. Data transmitted to earth from orbiting IR sensors can be viewed instantly on a large screen in normal light. G.E.

MEND
Medical Education for National Defense.

MER (Navy)
Proposal to send a man into orbit in a rocket-launched collapsible pneumatic glider. Rival plan to ADAM and Air Force project to put a man in space in 1959.

MERCURY (NASA)
Man-in-space program calling for nonorbital one-man flights with REDSTONE booster in 1960-61, and orbital one-man flights

with ATLAS booster in 1962. For the nonorbital shots, maximum altitude, about 15 miles; downrange distance, about 270 miles; time in flight, 10 to 15 min. Operation sequence: powered flight coasting period, drogue chute and main chute deployment, landing on the ocean, and recovery by helicopters. MERCURY-REDSTONE 2 carried chimpanzee Ham into a successful trajectory flight on 31 January 1961. MERCURY-REDSTONE 3 (weight, about 2 tons) carried Astronaut Navy Lieut, Comdr. Alan Shepard into a similar trajectory flight on 5 May 1961. MERCURY-REDSTONE 4 (weight, about 2 tons) carried Astronaut Air Force Capt. Virgil Grissom into a successful trajectory flight on 21 July 1961. First two flight capsules retrieved; third lost after Grissom's escape. MERCURY-ATLAS 5 launched on 29 November 1961, successfully orbiting chimpanzee Enos twice around the world before re-entry and recovery. MERCURY-ATLAS 6 to be first manned orbital flight, scheduled for January 1962. Thereafter, manned Mercury shots to be made every 60 days. MERCURY MARK 2 capsules to be used for two-man flights up to 14 days, with TITAN III booster. Diameter of MARK 2 capsule base, 83 in. Recovery weight, 5500 to 6000 lb. Recovery by either glide-sail chute, clustered small chutes, or 96-ft chute. Cost of first phase of MERCURY program may exceed $500 million. MARK 1 capsule cost per copy, about $2 million. Twenty-six capsules procured. Total cost of R&D and procurement for MERCURY program to end of 1961 of the order of $140 million. McDonnell, prime.

MERCURY II
See GEMINI.

MERCURY-SCOUT (NASA/Air Force)
Program of launching 100-lb satellites on a 300-mile orbit using SCOUT to test the world-wide MERCURY tracking network. First flight, summer 1961, with BLUE SCOUT booster. Ford-Aeronutronic.

METEO (U.S.S.R.)
Upper-air sounding rocket used to obtain meteorological data during the International Geophysical Year. Recoverable. Length, 23 ft. Weight, loaded, 1500 lb; payload, 159 lb; booster, 520 lb. Speed, 2450 mph. Altitude, 60 miles. Solid-propellant booster. Designed by the Soviet Academy of Science.

METEOR
Manned earth-satellite terminal evolving from earth-to-orbit

ferry rockets. Vehicle would be made of ferry rockets which would be taken apart after arriving in orbit and the parts re-assembled to build a giant space station. Proposal by Goodyear in the late 1940's.

METEOR (Air Force)
Long-range study of manned vehicle recovery systems applicable to airfields on the West Coast; also of the ground-based instrumentation needs of these systems.

METEOR (NASA)
Also known as D-58. Re-entry physics research vehicle. Speed, 24,000 to 160,000 mph. Six-stage system.

METEOR (Navy)
Air-to-air missile. Program initiated November 1945; canceled June 1954.

METEOR II
See SHAVIT SHTAYEM.

METHUSELAH
See X-7.

MET JET
Ramjet-powered meteorological sounding rocket. Solid propellant. Length, 98.5 in. Weight, 21 lb. Speed, over Mach 1. Capable of carrying 1.5-lb payload to a 200,000-ft altitude. Low cost, estimated at $400 per unit in quantity, based on a production rate of 100 per month. Tests have been made by Army Signal Missile Support Agency and by Navy meteorological group. Anderson, Greenwood & Co.

MHD
MagnetoHydroDynamics.

MIDAS (Air Force)
MIssile Defense Alarm System. Program for series of satellites in polar orbit to detect missile launches by infrared. First stage, ATLAS; second stage, AGENA B, which goes into orbit carrying the payload. Budgeted for approximately $105 million in FY 1961; $200 million in FY 1962.

MIDAS I
Launch made 26 February 1960. Satellite failed to enter orbit.

MIDAS II
Test vehicle launched successfully 24 May 1960. Apogee, 317 statute miles; perigee, 294 miles. Payload weight, 5000 lb. However, after a short time, transmitting devices failed.

MIDAS III
Satellite launched 12 July 1961. Entered polar circular orbit; apogee, 2197 miles; perigee, 2084 miles. Gross weight, 3500 lb.

MIDAS IV
Satellite launched 21 October 1961 into 2100-mile polar orbit. Carried piggyback experiment WEST FORD. IR detectors in MIDAS registered TITAN launching on 24 October 1961 within 2 min.

MIDGETMAN (Air Force)
Project to develop lightweight ICBM powered with a solid-fueled rocket engine. Also called SHRIMP and SICBM. Plans suspended. See also JOSHUA.

MIGHTYMAN (Air Force)
An ICBM system based on SATURN. In study phase.

MIGHTY MITE
See VANGUARD I.

MIGHTY MOUSE (Navy/Air Force)
Unguided missile, called the first modern artillery rocket. Weight, less than 20 lb. Carried explosive charge equal to a 75-mm shell. Similar in use, size, and appearance to FFAR. In production in 1951, and was deployed with Navy and Air Force interceptors. NOTS.

MILDDU (AIA)
Logistics program sponsored by Aerospace Industries Associates. Objective: a universal standard for military-industry support data interchange, reduction of spares overproduction, and improvement of speed and efficiency in supplying spares when needed.

MILS (Navy)
System of detection for impact of missiles in the ocean, particularly those launched from Vandenberg AFB. Two systems used: acoustic, and detection of a SOFAR (sound fixing and ranging) burst taking place under water.

MIND
Magnetic Integrator Neuron Duplicator. Electron "neurons" modeled after the synaptic junction of man's and animal's nervous tissue. Can remember experiences and learn new facts under control of human or mechanical teacher. Ford-Aeronutronic.

MINIAPS (AEC)
MINIature Accessory Power Supply. Solid-fueled APU. Application of an electrical turbine-driven generator in a self-contained power package (60-w). Sandia/Thompson Ramo Wooldridge/Tapco.

MINITRACK (NASA)
Radar tracking system established to track satellites. Cost, $4.75 million in FY 1961.

MINOTAUR
Boeing in-house small ICBM studies.

MINUTEMAN (Air Force)
Solid, three-stage ICBM, later generation than ATLAS and TITAN, to be launched from fixed bases. Military designations: SM-80; WS-133A. Length, 56 ft. Launch weight, 64,300 lb. Inertial guidance. Range 5500 nm. Speed, about 15,000 mph. Engines: XM55, XM56. First shot, 1 February 1961, was successful. Second shot, 19 May 1961, successful, but missile was destroyed because of guidance difficulties. Third shot, 27 July 1961, was successful. First launch attempt from silo, 30 August 1961, failed. MINUTEMAN bases are under construction near Great Falls, Mont.; Rapid City, S.D.; Minot, N.D.; and Sedalia, Mo. Also truck-launching systems under consideration. Possibility of 2000 MINUTEMAN missiles operational 1966-70. Boeing, prime; Autonetics, guidance; Thiokol, propulsion for first stage; Aerojet General, propulsion for second stage; Hercules Powder, propulsion for third stage; Avco, re-entry vehicle; AMF-ACF, rail launcher. MINUTEMAN cost per unit, about $3 million.

MIRF (Air Force)
Multiple Instantaneous Response File. Feasibility study for construction of a data-retrieval file which has large capacity and also permits simultaneous interrogation of all data. SRI.

MISS (Air Force)
   Man In Space Soonest. Project in 1958 led by Brig. Gen. Don
   Flickinger for a manned space capsule to be orbited by ATLAS-
   AGENA A in early 1960. Was to be followed by MISSOPH. Air
   Force interest in program waned when MERCURY program was
   assigned to NASA.

MISSILE A (Army)
   Surface-to-surface tactical support missile. Short range, 10 to
   20 miles. Launch weight, under 500 lb. Solid propellant. ARGMA,
   Huntsville, Ala., acting as systems manager for components
   under development. With MISSILE B, to be possible replace-
   ments for LITTLE JOHN, HONEST JOHN, and LACROSSE.

MISSILE B (Army)
   Tactical support missile. Comparatively lightweight. Possible
   range, 70 miles. Solid propellant. Project combined with MISSILE
   A. Design studies by G.E., Cornell Aeronautical Laboratory,
   Martin, Armour, Minneapolis-Honeywell, and Douglas.

MISSILE C (Army)
   Surface-to-surface missile. Range, 70 to 90 miles. Planned
   follow-on to SERGEANT.

MISSILE D (Army)
   Tactical missile. Range, over 500 miles. Might be an improved
   PERSHING. In study phase.

MISSILEER (Navy)
   Aircraft intended to carry the EAGLE missile. Program suspended. Douglas.

MISSILE MASTER (Army)
   An electronic system for receiving early-warning information.
   Military designation, MSG-4. Achieves maximum effectiveness
   in firing all antiaircraft weapons at any given location by coordinating all parts of antiaircraft defense. Martin. See BIRDIE.

MISSOPH (Air Force)
   Man In Space SOPHisticated. Project, led by Brig. Gen. Don
   Flickinger, to have followed MISS in the 1961-63 time period.
   Consisted of the orbiting of a 6000-lb manned capsule, boosted
   by ATLAS-AGENA B, and a 9000-lb capsule boosted by ATLAS-
   CENTAUR. Project not pursued.

MISTRAM (Air Firce)
MISsile TRAjectory Measurement system. Purpose: to provide primary range safety and instrumentation on all missile flights, and of Cape Canaveral. Intended for Atlantic Missile Range. Under development by G.E. First operational tests scheduled for 1962.

M.I.T.
Massachusetts Institute of Technology, Cambridge, Mass.

MITRE CORP.
Formed by Air Force in September 1958 to support Air Defense Systems Integration Division (ADSID) technically. Expanded to about 1500 personnel (of whom 500 are professional). Contracts for FY 1962 expected to be of the order of $30 million. Specializes in electronic command and controls systems.

MM.20 (France)
Surface-to-surface guided missile developed by Nord Aviation from the CT.20 jet target drone. Designed for shipboard use against naval targets. Range, approximately 155 miles. Warhead capability, 550 lb. See also SM.20, R.20.

MMRBM
Mobile Medium-Range Ballistic Missile.

MOBIDIC (Army)
Army's first operational mobile digital computer. Completely transistorized. Controls Seventh Army supply requisitions for equipment items. R&D by Sylvania/Signal Corps.

MOBILITY (Army)
Army exercise at Fort Eustis, Va., demonstrating an experimental rocket lift device which enables the wearer to achieve controlled free flight over the ground. See SRLD.

MOBOT II
MObile roBOT. A mobile manipulating system, equipped with TV "eyes" for remote control. Electromechanical. Proposed for use in exploring the surface of the moon. Hughes.

MODEL NO. 30 (Japan)
Surface-to-surface rocket under development by Japanese Defense Agency. Length, 4.5 m. Diameter, 30 cm. Solid propellant. Unguided. Similar to U.S. LITTLE JOHN. To be

operational in 1966. Planned for deployment with one battalion in the Ground Self-Defense Force.

MOHO (Army/National Science Foundation)
Also MOHOLE. Project to drill through the ocean floor to the Mohorovicic layer of discontinuity between the earth's crust and mantle. Tests have been made off the west coast of Mexico.

MONICA (France)
Three-stage, upper-atmosphere research rocket. Five models made, with varying altitude, payload, and performance capabilities. Solid propellant. Length, varies from 9.8 ft to 20.5 ft. Weight, 154 lb. to 340 lb. Altitude, 23 to 100 miles. Velocity, 1700 to 3600 mph. Cost, about $1500, without instrumentation. Developed by the Association Technique pour l'Etude des Fusées under contract of the Ministère de l'Air.

MOONBEAM
Proposed lunar system. Probably a five-stage chemical rocket (such as a SATURN configuration) or a four-stage chemical booster with nuclear upper stages. G. D. — Astronautics.

MOON-CRAWLER
Vehicle proposed by RCA to gather scientific information on the moon. Proposed weight, 3000 lb. To be capable of moving about on the moon in response to radio instructions from earth. Unmanned. Outfitted with a TV "eye," a claw for picking up surface rock and dust samples, and antennas for communication with earth bases. Launching could be by SATURN booster. Remote control to be used to slow vehicle on approaching the moon.

MOON-HUT
Self-sustaining living unit proposed by Martin in 1960, to be the permanent moon base for first astronauts.

MOONLIGHT PHOTODETECTOR (NASA)
Orientation device capable of sensing weak rays from the moon and of withstanding strong rays from the sun. Designed for space instruments. IBM.

MOONSHOT (NASA)
Lunar orbiter with a photographic system, 100-m resolution, to man the moon. ATLAS-CENTAUR booster. Scheduled launch, 1964. Two shots planned. Study contracts to Eastman, Fairchild, and RCA.

MOSQUITO (Switzerland - Army)
Surface-to-surface antitank missile. Solid propellant. Length, 3 ft. Launch weight, 23 lb. Wire guidance. 7-lb HE warhead. Range, about 6200 ft. Speed, 200 mph. Deployed with Swiss Army units. Operational. Contraves AG/Oerlikon/Buhrle.

MOTH (National Defense Research Committee)
Glide bomb. Variation of the PELICAN. Designed to home in on enemy radar.

MOUSE
Minimum Orbital Unmanned SatellitE. A 100-lb satellite with limited life proposed by S. Fred Singer in 1954. Never built. VANGUARD and EXPLORER were influenced by many of its design features.

MR-1 to MR-8 (NASA)
Redstone-boosted MERCURY capsules in the Man-in-Space program. McDonnell.

MRBM
MidRange (or Medium-Range) Ballistic Missile.

MRBM (Air Force)
Medium Range Ballistic Missile. A mobile missile system capable of transportation by truck-launchers. Could carry nuclear warhead. Under consideration. No contract but Boeing and Convair are expected to be strong contenders.

MRICC (Army)
Missile and Rocket Inventory Control Center. At U.S. Army Ordnance Missile Command, Huntsville, Ala.

MRS. V (DOD)
Maneuverable Recoverable Space Vehicle. Outgrowth of MERCURY plus DISCOVERER and DYNA-SOAR programs. Concept that pilot could change his orbit and speed, assemble space stations, and intercept other space vehicles. In R&D.

MSG-4 (Army)
See MISSILE MASTER.

MT
Machine Translation.

MTG-CI (Italy – Air Force)
Beam-riding, surface-to-air weapon system. Military designation, RSD-58.

MTSS (Air Force)
Military Test Space Station. Military designation, SR-17527. An Air Force study in 1961. Lockheed Missiles and Space.

MU (Japan)
Reported to be a large scientific rocket capable of reaching 2600-mile altitudes. To be successor to LAMBDA. Under study.

MUROC (Air Force)
Surface-to-surface missile in R&D.

MUSCLEMAN (Air Force)
An ICBM, successor to MINUTEMAN, with increased range and payload. In planning stage. See also MIDGETMAN.

MUSKRAT (Navy)
Naval Research Laboratory cosmic-ray research program.

MX
Missile Experimental. Code numbers follow to designate experimental missile project.

MX-77A
See SHRIKE.

MX-570
See THIAMAT.

MX-770
See NAVAHO, NATIV.

MX-771
See MATADOR.

MX-772 (Air Force)
Surface-to-surface missile. Program begun March 1946; canceled June 1947.

MX-773 (Air Force)
Surface-to-surface missile. Program begun February 1946; canceled May 1947.

MX-774 (Air Force)
Surface-to-surface missile. Program begun May 1946; canceled
June 1947.

MX-775
See SNARK.

MX-776
See RASCAL.

MX-904
See FALCON.

MX-1593 (Air Force)
Surface-to-air missile. Program begun 1951; canceled 1954.
See ATLAS.

MX-2013 (Air Force/Navy)
Air-to-surface missile designed to home on enemy radar antenna. Radioplane.

N

NA-273
See REDHEAD/ROADRUNNER.

NABS
Satellite in R&D. No further information available.

NACA
National Advisory Committee for Aeronautics. Reorganized to
become NASA.

NACCAM
NAtional Coordinating Committee for Aviation Meteorology.
Consists of heads of meteorological services of the three military services, plus FAA, NASA, and U.S. Weather Bureau
representatives. See POMS.

NAKA (Air Force)
Air-to-air missile. Length, 24 in. Width, 1.5 in. North American project before 1957. Rocketdyne.

NANWEP (Navy)
Research project to develop weather analyses and forecasts for

special Navy requirements by the use of high-speed computers. Project's first operation product transmitted to PMR on a test basis early in 1961.

NAP-100 (Air Force)

Thermoelectric generator using solid radioactive isotope, Curium 242. Weight, 40 lb. Operates continuously for a year.

NASA

National Aeronautics and Space Administration. Formerly NACA. A civilian agency in control of the greater portion of American space research.

NASTY (Air Force)

Air-to-air missile. Intended as a deterrent to enemy fighters. Phased out. NAA.

NATIV (Air Force)

North American Test Instrument Vehicle. Military designations: RTV-A-3; MX-770. Test vehicle for NAVAHO missile. Designed to check out aspects of guided missile flight and performance. Length, 13.25 ft. Weight, loaded, 1100 to 1250 lb; empty, 610 lb. Range, 40 to 60 miles. Altitude, 12 miles. Speed, Mach 3. Liquid propellant. First firing in the summer of 1948, with a 32-channel telemetering system which transmitted to earth the measurements made on the vehicle. NAA.

NATO

North Atlantic Treaty Organization. In French, OTAN (Organisation du Traité de l'Atlantique du Nord).

NAVAHO (Air Force)

Surface-to-surface, ramjet-powered missile. Military designations: SM-64; MX-770; XSSM-A-2. Length, 76 ft. Weight, 100,000 lb. Inertial, celestial, IR homing guidance. Liquid propellant. Speed, Mach 3. Range, 5000 miles. Thrust, 275,000 to 400,000 lb. Program begun March 1946, canceled in July 1957 after expenditure of approximately $680 million. Vehicles in existence after the cancellation were used in research, and the REDSTONE, JUPITER, THOR, and ATLAS engines stem from it. NAA.

NAVSEA (Navy)

Naval AVionics Support Equipment Appraisal.

NAVSPASUR (Navy)

NAVy SPAce SURveillance system. Also SPASUR. Purpose: to

detect and determine the orbits of all nonradiating space objects within its range. Operational since 1959.

NBS
   National Bureau of Standards.

NBSR (NBS)
   High flux research reactor to be built at Gaithersburg, Md. Purpose: to provide neutron and gamma-ray sources for NBS and other governmental agencies; also to provide capability for the analysis of atomic and molecular structure by neutron diffraction.

NCAR (NSF)
   National Center for Atmospheric Research. Established in June 1960 by the National Science Foundation and a managing corporation representing 14 U.S. universities. Purpose: to study the effects on weather and climate from all possible influences, e.g., lightning, sun, electricity, oceans, forests, and deserts. Programs to be handled by American and foreign scientists. Also some projects to be contracted among universities and other laboratories.

NEEDLES
   Former name for Project WEST FORD.

NEKTON II
   Phase of TRIESTE program.

NELL
   Name given series of rockets developed by Dr. Robert N. Goddard between 1914 and 1945. These rockets considered the inspiration for the German V-2's and the forerunners of modern guided missiles.

NEPA
   Study of use of atomic propulsion for aircraft, initiated 28 May 1946.

NEPTUNE
   See VIKING.

NERV (NASA)
   Nuclear Emulsion Recovery Vehicle. Designation, P-26. Mission: to probe Van Allen radiation belts and gather basic data needed for development of protection of space crews from radia-

tion. Boosted by ARGO D-8, 19 September 1960. Recovered. See also BIOS I.

NERVA (AEC - NASA)
Nuclear Engine for Rocket Vehicle Applications. Follow-on development to KIWI B. The first flyable ROVER engine. May be test-flown by 1965.

NEW HORIZONS
Experimental missile, in R&D.

NIKE-AJAX (Army)
First American operational surface-to-air missile. Military designations: SAM-A-7; XSAM-A-7. Length, 21 ft; with booster, 34 ft. Weight, 2300 lb. Liquid-solid propellant. Speed, Mach 2. Range, 25 to 30 miles. Engine: See X-213. Phased out of production and being replaced by NIKE-HERCULES. Deployed in U.S., Europe, and Far East. Western Electric, prime, guidance; Thiokol, propulsion; Douglas, airframe.

NIKE-APACHE (Army)
Research vehicle. To boost 50- to 150-lb payload. In R&D. New Mexico State College.

NIKE-ASP (Army)
Also called ASPAN. Solid-propellant research vehicle capable of lofting a 25-lb payload to a 150-mile altitude. Operational. Marquardt – Cooper.

NIKE-CAJUN (Air Force/NASA)
Research vehicle capable of lofting a 50-lb payload to a 100-mile altitude. Operational; several hundred flown. Various manufacturers, including Marquardt-Cooper.

NIKE-HERCULES (Army)
Surface-to-air antiaircraft missile. Military designations: SAM-A-25; XSAM-A-25. Length, 39 ft. Weight, 10,000 lb. Beam-rider command guidance; semiactive homing. Solid propellant. Speed, 4000 mph. Range, 50 to 75 miles. Thrust, 2600 lb. An advanced NIKE, with both nuclear and conventional warhead capability. Operational in 1958. Replacing NIKE-AJAX installations. Engine: See XM30. Western Electric, prime, guidance; Hercules/Thiokol, propulsion; Douglas, airframe.

NIKE-NIKE (Air Force)
Five-stage research vehicle capable of lofting a 50-lb payload

to a 100-mile altitude. Solid propellant. Operational; no production. Marquardt-Cooper.

NIKE-VIPER
High altitude sampler rocket (HAS). See TUMBLEWEED.

NIKE-ZEUS (Army)
Surface-to-air, anti-ICBM and -IRBM, three-stage missile. Solid propellant. Range, 200+ miles. Nuclear. Weight, 40,000 lb. Length, about 65 ft. Speed, over Mach 4. Engine: TX-135; XM6. Only R & D funding to date. Production not yet authorized. See ZEMAR. Western Electric, prime; Douglas, airframe, GSE; Hercules/Thiokol, propulsion.

NIMBUS (NASA)
Polar-orbiting meteorological satellite. Weight, 650 lb. Follow-on to TIROS series. To be launched into 600-mile orbit with THOR-AGENA B booster. Instrumentation to include RCA vidicon cameras for cloud data, radiation sensors at various resolutions, and spectral ranges for heat balance, temperature, and cloud cover. Also to include stabilization system to keep camera pointed toward earth. First launching in series scheduled for mid-1962. Budgeted, with TIROS, for $50 million, FY 1962. G.E., prime.

NITRASOL
High-energy propellant developed by Lockheed Propulsion (formerly Grand Central Rocket). Will not detonate when hit by small-arms fire, thus usable by military troops under field conditions.

NMSSA
NATO Maintenance Supply Services Agency. Established in April 1958 to manage logistic operations for NATO.

NNRDC (AEC)
National Nuclear Rocket Development Center. To be established in Nevada.

NODAC (Navy)
Naval Ordnance Data Automation Center. A data-reduction machine adaptable in a very short time to any data-reduction problem by means of plug-in logic packages and patchboards.

NOMAD (Air Force/NASA)
Upper-stage fluorine rocket engine with 15,000-lb thrust. Ca-

pable of landing a 1200-lb payload on the moon. Designed for use with ATLAS. In research. Bell Aerosystems.

NOMAD I (NBS/Dept. of Commerce/BuWeps)
A robot weather station consisting of a platform measuring approximately 10 x 20 ft, with four airtight aluminum wells for weather-recording gear.

NORAD (Air Force)
NORth American Air Defense Command. Underground combat operations center near Colorado Springs, Colo. Fully operational in late 1961.

NORC (Naval Weapons Laboratory)
Naval Ordnance Research Calculator. An "electronic brain" which operates at a capacity of 15,000 operations per second, 13-decimal digit numbers.

NORTOBRAZE (Air Force)
Process using electronically controlled banks of radiant quartz lamps to braze stainless-steel honeycomb panels. Developed for use in manufacturing advanced aircraft and space vehicles. Northrop.

NOTS (Navy)
Low-cost satellite launcher aircraft. Supersonic.

NOTS (Navy)
Naval Ordnance Test Station. Located at China Lake, Calif.

NOTSNIK (Navy)
Name given to project calling for launch of small satellites from aircraft. A four-stage missile (PILOT) was used as booster in 1958 but did not attain orbital flight. Program has been continued under Project CALEB. NOTS.

NOTUS (Army)
Communication satellite project which includes former projects COURIER, STEER, and DECREE. Reoriented to include only COURIER and Navy ADVENT. In R&D phase. See also COURIER, ADVENT.

NOVA (NASA)
Program for a big booster beyond SATURN. Stages and propulsion system not as yet firmed up. A SATURN C-5, using five F-1 engines in the first stage to provide more than half the performance desired in the NOVA class booster. Air Force PHOENIX studies may result in big booster configurations, some

using solids, especially for the first stage. Air Force seeking a 120-in.-diameter solid booster which could be clustered. Such a booster is scheduled for development as part of TITAN III. Success in rendezvous will determine the importance of the NOVA program. See NOVA-L.

NOVA-L (NASA)

A configuration planned as a parallel program to SATURN C-5. SATURN is designed for the redezvous approach to lunar landing and NOVA for the direct assault. If the rendezvous method proves successful, the NOVA vehicle might not be needed for lunar missions. The presently planned NASA NOVA is liquid. To be capable of boosting 400,000 lb into a 300-mile orbit, or 150,000 lb to the moon. Stage 1 (S-1D): eight Rocketdyne F-1 engines (LOX-RP); total thrust of stage, 12 million lb; budget for FY 1963, $55 million; total budget presently planned, $163.5 million. Stage 2 (S-2B): four Aerojet M-1 engines (LOX-H_2); total thrust, 418 million lb. Stage 3 (S-3B): one Aerojet M-1 engine; thrust, 1.2 million lb; budget for FY 1962, $26 million; for FY 1963, $55 million; total budget presently planned, $163.1 million. No contracts awarded for stages (January 1962).

NOVA-S (Air Force)

Concept of a clustered 12-million-lb solid-propellant booster. Diameter desired, about 120 in. for each unit in cluster. Parallel program to NOVA-L and possible alternative. If such a booster is developed, it may assume a new name. Under study. See also PHOENIX.

NOVUS

Proposal for a 14.4 million-lb-thrust solid-propellant booster made by Lockheed Propulsion. Booster to be composed of six clustered 2.4 million-lb-thrust engines. Capable of orbiting 150,000-lb payload at a 300-mile altitude.

NRL

Naval Research Laboratory.

NSF

National Science Foundation.

NUDETS (Air Force)

NUclear DETection System. Military designation, WS-477L. Purpose: to locate and measure the yield of nuclear blasts. System to consist of sensors and computers which will correlate the information. To be located at strategic points in the U.S. Planned installation and operation date is early 1963.

NULACE (Air Force)

NUclear Liquid Air Cycle Engine. Nuclear version of LACE. I_{sp} said to be twice that of the ROVER nuclear rocket engine. Under development by Marquardt for possible Aerospace Plane application.

NX-2

G. D.—Astronautics proposal for a nuclear-powered aircraft.

OAO (NASA)

Orbiting Astronomical Observatory series. Three satellites scheduled; more planned. Weight of satellites, approximately 3300 lb, including 1000 lb of experiments. Total cost estimated at about $100 million.

OAO I

Octagonal-shaped standard shell of aluminum, 9.5 ft high by 6.5 ft wide, with 40-in.-diameter experiment chamber. To have stabilizing system to lock astronomical equipment on celestial body under observation. Satellite command system to receive ground signals to point and operate the satellite and its experiments. TV tube to transmit pictures to ground stations. Experiments: several 8-in. telescopes connected to TV tube to map ultraviolet radiation of entire sky (Smithsonian Astrophysical Observatory); instruments to measure brightness of ultraviolet emissions from the stars (University of Wisconsin); 36-in. mirror and spectrometer to study emissions from great number of heavenly bodies (Goddard Space Flight Center); spectrograph to make studies of sun's activities (Harvard University); 24-in. mirror and spectrometer to study cosmic gas and dust by observing them against the stars (Princeton University Observatory). Scheduled for launch into 550-mile orbit from Cape Canaveral by ATLAS-AGENA B booster in spring of 1963. Grumman, prime; Westinghouse, electronic components; E.G., stabilization and control.

OAO II

Satellite to contain experiments designed by Goddard Space Flight Center. Purpose: to obtain absolute spectrophotometric data on selected stars, nebulae, and galaxies. Optical system to employ relatively fast 36-in. Cassegrain telescope with a large-aperture spectrophotometer. Scheduled for launch 1964. Grumman, prime.

OAO III
Satellite to contain experiments devised by Princeton University to provide quantitative observations of the absorption spectrum of the interstellar gas in the regions 800 to 1500 A and 1600 to 3000 A. Scheduled for launch in late 1965.

OAO IV
This and later satellites expected to be used for studies of the sun and planets.

OAR
See AFOAR.

OGO (NASA)
Orbiting Geophysical Measurement Observatory. Program to loft 1000-lb satellites equipment with instruments for geophysical measurements into polar (POGO) and eccentric (EGO) orbits. Satellite capable of carrying instruments for 50 different experiments on any one mission. Length, 6 ft. Diameter, 3 ft. Boosters, ATLAS-AGENA B, THOR-AGENA B, and CENTAUR. First flight scheduled for 1963. Total cost for program estimated at about $150 million. Cost per unit, about $5 million. STL.

OILSAND (AEC)
Project proposed to release trapped oil deposits in Athabaska area of Alberta, Canada, by use of atomic blasts underground. Part of PLOWSHARE.

OLO (NASA)
Orbital Launch Operations. Study of all phases of the orbital launch mission, including equipment, crew requirements, program scheduling, and costs breakdown. Six-months' study contract to Ling-Temco Vought-Astronautics.

OMAR (Navy)
Air-to-surface HVAR-type missile never carried beyond development stage. Military designation, XASM-N-6. Length, 10 ft. Diameter, 5 in. May have been optically guided. Canceled in October 1954. Eastman Kodak/Armour Research Foundation.

OMEGA (Navy)
World-wide navigation system. BuShips.

OMICRON BELT
Term used for the orbiting fragments of the ABLE STAR rocket which lofted TRANSIT IVA and the piggyback satellites INJUN and GREB on 29 June 1961. When the rocket exploded in orbit, 42 fragments were scattered in a cloud of debris which is a complication for the SPACEWARN network, and which may also be a hazard for future space vehicles.

OMNIFORM I (NBS)
General-purpose program enabling scientists to apply high-speed computation methods directly to research problems. Program allows even users who are unfamiliar with computer programing techniques to instruct the machine to perform specialized calculations. Primarily a table generator.

OMSA (Army)
Army Ordnance Missile Support Agency. Formerly the Redstone Arsenal element of AOMC. Purpose: to provide support services for ABMA, Hq., AOMC, the Guided Missile School, and the George C. Marshall Space Flight Center; also to provide new services, as a missile and rocket inventory control center a computation center, calibration laboratory, and an Army missile patent center.

ONERA
Office Nationale d'Etudes et de Recherches Aéronautiques National Office of Aeronautical Study and Research.

ONR (Navy)
Office of Naval Research.

OPERA (Army)
Ordnance Pulses Experimental Research Assembly. Special nuclear reactor producing pulses of radiation. Preliminary work being done by Aerojet-General-Nucleonics with Aberdeen Proving Ground.

OPTAG I
Optical Pickoff Two Axis Gyro. Extremely small null-sensing autocollimator. Two-axis. Eliminates inaccuracies due to standard electrical precession pickoff by use of the principle of autocollimation. Company-sponsored program by Perkin-Elmer.

OPTEVFOR (Navy)
OPerational Test and EValuation FORce.

OQ-19; OQ-19D, E (Army)
Target drone, principally of aluminum, used with NIKE. Length, 12.25 ft. Weight, about 325 lb. Speed, 225 mph. Altitude, 4.4 miles. Liquid propellant. Same vehicle as Air Force OQ-19B and Navy KD2R-5. Advanced versions are the RP-87 and RP-88. See OQ-19B.

OQ-19B (Air Force)
Target drone. Same vehicle as Army OQ-19, OQ-19D, E, and Navy KD2R-5.

ORANGE (Air Force)
See TEAK.

ORBITER
Army—von Braun proposal in 1954 to get a satellite into orbit using a REDSTONE booster. In 1958 the Army orbited EXPLORER I using ORBITER concepts. ORBITER program discontinued in 1955 when the VANGUARD project was established.

ORCON (Navy)
ORganic CONtrol. Attempt to use pigeons as homing devices in missile noses. Original study by the University of Minnesota during World War II as means of neutralizing magnetic, thermal, and acoustic homing mechanisms. Reported to have worked well but no actual flight tests made. Project abandoned.

ORDIR
Triangulation of a group of radars for accuracy in locating objects in space. See ESAR, PINCUSHION, DEFENDER.

ORIOLE (Navy)
Air-to-air missile. Solid propellant. Military designation, XAAM-N-4. Weight, approximately 400 lb. Range, 5 to 10 miles. Speed, about Mach 2. Program began May 1948; canceled January 1954 in favor of SPARROW. Martin.

ORION (Air Force)
Formerly an ARPA project for pulsed-thrust system propulsion. First of the nuclear-pulse systems. ORION I to be capable of lofting 1000-ton satellite into orbit or similar-weight platform into space. Program begun in 1958 with a $2.5 million feasibility study contract to G.D.—General Atomics. ORION II to have capability for lofting up to 20,000 tons into space. Would permit establishment of large lunar bases, exploration of the solar system, and cislunar orbits. Martin proposed three types of

nuclear-pulse rockets different from the ORION concept by boosting the vehicle to a 150-mile altitude by a chemical rocket cluster before starting the nuclear-bomb propulsion. ORION feasibility tests made by Project PUT-PUT, but with HE charges simulating the nuclear explosions. No nuclear trials made because of the international moratorium. A flight-test vehicle could be developed within 5 years. Total development costs for an operational vehicle estimated at $1.5 billion.

ORION (Air Force)
Airborne medium-stage ballistic missile.

ORIONE SAR (Italy)
Winged surface-to-air weapon. Limited range. Stacchini.

OSCAR (Air Force)
Orbiting Satellites Carrying Amateur Radios. Project OSCAR club founded by Lockheed scientists. A project to enable ham radio fans to acquire and track satellites. First OSCAR transmitter a box-shaped, 10-lb satellite. Piggyback on DISCOVERER XXXVI shot on 12 December 1961. Temperature information transmitted by signals (145 mc) that can be picked up by ham receivers. Transmitted four dots and two dots, or "HI." Lockheed.

OSO (NASA)
Orbiting Solar Observatory. Designations, S-16, S-16A, S-17. Weight of satellites, 350 lb. S-16 launched into circular orbit by THOR-DELTA booster, 2 October 1961, for purpose of measuring solar phenomena. S-16A, scheduled for launch February 1962, designed to make maximum use of payload volume of THOR-DELTA vehicle. Primary objective of S-16A: to collect information of sun spectra in the ultraviolet and X-ray bands. Over-all height, 37 in. Total weight, about 440 lb. Experiments contained: high-resolution Lyman-Alpha spectrometer (University of Colorado); ultraviolet and X-ray spectrometer and low-energy gamma-ray monitor (Goddard Space Flight Center); soft gamma-ray solar monitor (University of Minnesota); hard gamma-ray solar monitor (University of Rochester); materials experiment (Ames Research Center); neutron flux sensor and electron-proton analyzer (University of California). S-17 to be instrumented similar to S-16A, with one modification. On command, the pointing instruments will scan the sun, giving a picture very like TV. Cost per satellite estimated at $8 million. Ball Bros., prime.

OSPREY (Army)
Surveillance drone with a miniaturized inertial guidance system. Military designations: AN/USD-5; USD-5; SD-5. Weight, 10,000 lb. Length, 36 ft. Delta wing configuration. Fairchild.

OSR
See AFOSR.

OTS
Office of Technical Services, U.S. Dept. of Commerce.

OUTPOST I, II
Two four-man space laboratories to be orbited at 300-mile altitude. Men to be ferried to satellite by glide rockets. CENTAUR booster. OUTPOST I would be the empty ATLAS oxygen tank, equipped with laboratory and capable of sustaining the crew. A 1959 proposal by Convair.

OYSTER
General-purpose underwater transducer, for use in submarine detection. Clevite Corp.

OWL (Air Force)
Research vehicle developed in 1959 and used to determine wind direction and velocity to 240,000-ft altitudes. Weight, 29 lb. Solid-propellant rocket boosts a small ballistic dart which carries a timer set to eject radar-reflective instrumentation at indicated altitudes. Successfully tested at Holloman AFB. Borg-Warner (BJ Electronics).

OZARC (ONR)
OZone ARCas. Study of air circulation by firing upper air ozone sampling rockets (ARCAS) simultaneously from several widely separated positions in the Pacific.

OZMA (ARPA)
Attempts to pick up signals possibly beamed from intelligent beings in space by use of an 85-ft radio telescope at Green Bank, W. Va. After unsuccessful trial in 1960, program temporarily discontinued. May be resumed with larger radio telescopes either at Green Bank or in Puerto Rico.

P

P-14
See EXPLORER X.

P-21, 21A (NASA)
Two geophysical probes to study electron distribution in the ionosphere. SCOUT booster. Launch date, late 1961.

P-26
See NERV.

P-31 (NASA)
Lunar orbiter, ATLAS-ABLE booster. Weight, 400 lb. Failed 1960.

P-32, 33, 34, 35, 36
See RANGER.

P-46
See EXPLORER VIII.

P-60-1A, 2
See PIONEER.

PACE
Industrial time study evaluation program.

PACIR (U.S. Patent Office)
A data-retrieval system developed by IBM. Uses a broad chemical-compound search system.

PADAR (FAA/Air Force)
PAssive Detection And Ranging. An anticollision device. Fairchild Astrionics.

PADDLEWHEEL (NASA)
Term applied to the spacecraft of EXPLORER VI (also known as ABLE III) because of its shape—an irregular spheroid with a somewhat flattened lower part, with four solar-cell paddles protruding from the vehicle. Total weight, about 142 lb. Spheroid diameter, 26 in. Diameter with extended paddles, 86 in. Height, including interstage structure, 29 in. Solar paddles, 20 x 22 in. on the surface. PADDLEWHEEL also a general term for spacecraft of similar design.

PANDORA (Great Britain - RAF)
Reportedly an air-to-surface missile under development for the RAF. Called a "flying bomb." Designed for use with TSR-2

planes which are to have capability of flying at transonic speed at treetop altitudes.

PANGLOSS (Navy)
Study of means for secure communications between POLARIS missile submarines and aircraft surface ships and other submarines. RCA, David Sarnoff Div.

PARA (Air Force/Navy)
Study of memory systems and associated logic circuits. In R&D by Cornell Aeronautical Laboratory. Sponsored jointly by ONR and Air Force (Rome ADC).

PARADE (Army)
Passive-Active RAnge DEtermination. A classified electronic countermeasures system. Prototype contract to Sylvania in 1960.

PARCA (France - Army)
Surface-to-air missile. Solid propellant. Length, 18 ft. Launch weight, 2200 lb. Range, 15 miles. Speed, Mach 1.5. Used mainly for training French combat troops. Advanced version under development for use against Mach 3 planes. DEFA.

PARSECS
Program for Astronomical Research and Scientific Experiments Concerning Space. Long-range company-funded research program by Boeing, designed to keep the company in advance of present U.S. space-exploration activities. Main activities include manned earth satellite observatory; moon colony; countermoon (an earth satellite at lunar distance and on opposite side of the earth to the moon); interplanetary probes; close solar orbit; Trojan-point observatories (observation posts 60° ahead and 60° behind earth in same orbit around the sun); out-of-ecliptic orbit; and planetary exploration.

PAT-1 (Argentina - Air Force)
Air-to-surface liquid-propellant glide bomb. Range, 12 miles. Length, about 11 ft. Launch weight, 2310 lb. Speed, 560 mph. Based on German World War II design. Developed by Instituto de Investigaciones Tecnicas.

PCH (Navy)
Patrol Craft Hydrofoil. Experimental hydrofoil craft designed for antisubmarine warfare mission. $2 million contract to Boeing–Aerospace.

PD-121 (Army)
Drone with cruise range of 230 miles. Speed, Mach 3. Used to train missile crews. Beech.

PD-143 (Army)
Target drone used for training missile personnel and evaluating new weapon systems. Maximum speed, 690 mph at sea level. Beech.

PEEPING TOM
See SNOOPER.

PEGASE (France)
High-altitude sounding rocket composed of BELIER rocket as upper stage fitted on two-stage booster. Will loft 40-lb payload to a 620-mile altitude. Speed, Mach 12.5. Unguided. Solid propellant. One of a family of inexpensive rockets being developed by Sud-Aviation. See BELIER, CENTAURE, AIGLE, DRAGON, ERIDAN.

PEGASUS C
Modified version of X-17 proposed 1958, to improve performance and lower cost. Not developed. Lockheed Missiles and Space. See X-17.

PELICAN (National Defense Research Committee)
Early test vehicle (glide bomb) for the BAT antiship program. In limited production in 1944. One version capable of carrying a 500-lb payload. Another version could carry a 1000-lb payload.

PENCIL
One-, two-, and three-stage rockets less than 1 ft long designed to study the relationship between center-of-gravity location and dispersion. Also simulated larger rockets and trained personnel in launching techniques. Probably smallest ever used for test information. Over 200 fired in 1955. Cost, approximately $15 each. Produced by University of Tokyo.

PENNY
Air-launched, supersonic, expendable target. Dual-chamber, liquid rocket engine. Bendix, Bell Aircraft. See HILO.

PEP (Air Force)
Management technique similar to Navy PERT. Designed to prevent slippage of programs or deterioration of quality assurance. Used on B-70 and GAM-87 programs.

PERCEPTRON
A learning machine which learns initially from a human teacher, as does the Raytheon CYBERTRON. However, the PERCEPTRON is an attempt to duplicate the neural networks of the brain by electronics rather than by the human learning process. Cornell Aeronautical Laboratory.

PERCHERON
Sounding rocket composed of one POLLUX and two RECRUITS. Has capability of lofting a 2200-lb payload to a 50-mile altitude. One flown for Air Force in September 1960. Aerolab.

PERSHING (Army)
Surface-to-surface, two-stage field artillery missile. Length, 34 ft. Solid propellant. Inertial guidance. Developed as tactical replacement for REDSTONE. Range, 350 nm. Supersonic speed. Capable of greatly increased range. Operational in late 1961. Under consideration by NATO. Martin, prime; Bendix, guidance; Thiokol, propulsion. See MISSILE D.

PERT (Navy)
Program Evaluation and Review Technique. A program to monitor and control a major project through development and production, using data-processing and computer techniques, in order to maintain optimum schedule and reduce costs.

PERT II (Air Force)
Advanced PERT management system for MINUTEMAN. Contract to operations Research, Inc., Western Div. Improvement over PERT is "management by exception" techniques and improved automation.

PET
Tiny control rocket used to shut valves, stop machinery, or build up pressure quickly in satellites or spacecraft. Length, 4.8 in. Atlantic Research.

PETREL (Navy)
Air-to-surface, "bat"-type antisubmarine missile. Length, 24 ft. Weight, 3800 lb. Short range. Turbojet-powered. Radar-homing guidance. Thrust, 1000 lb. Program begun in August 1944; terminated in July 1957. Engine: See J-44. Fairchild/U.S. Bureau of Standards.

PHOENIX (Air Force)
Two-stage research rocket capable of lofting a 10-lb payload to 250-mile altitude. Length, 18 ft. Weight, 325 lb. Firings were made at the Pacific Missile Range for the University of Maryland under contract with Cambridge Research Laboratories. Also fired at Wallops Island (NASA) for the Air Force by the Leesona Corp. Rocket Power/Talco.

PHOENIX (Air Force)
A series of studies aimed to resolve the large booster problem for the Air Force. Most propulsion companies and several aerospace companies have participated in these studies. Solid propellants especially are under consideration, but liquids are not excluded. Various staging and recovery techniques, configurations, and unorthodox launching methods considered. One configuration mentioned is two-stage having a half-million-lb thrust solid first stage and a 200,000-lb liquid hydrogen-oxygen second stage. The PHOENIX could lead to an alternate booster to SATURN and for NOVA. Aerospace/RAND leading the study effort for the Air Force. See also PROJECT 3059.

PHOTOPAT (Army)
A photoelectric semiconductor device for use in guidance systems. Measures and reports electronically various types of motion in missiles or spacecraft. In development stage. Giannini Controls.

PI (Japan)
See SIGMA (Japan).

PICE
Two high-speed data-processing systems developed for Lockheed under a $1.1 million contract. To control data flow between earth and satellites. Beckman Instruments.

PIED PIPER
See DISCOVERER and related programs.

PIGGYBACK
Classified project. Purpose: to carry environment measuring packages under the Advanced Information Collection program. See also SPACER.

PILAC
Pulsed Ion Linear Accelerator. System incorporates a compression magnet. Converts a dc current of high-energy ions

into a series of intense, equally spaced bursts, each but a fraction of 10^{-9} second long. Los Alamos Scientific Lab.

PILGRIM
G. E. proposal for a lunar colony to be established in the late 1960's. To use the SATURN booster. Estimated mission cost, $8 billion.

PILLBOX
See TIROS I.

PILOT (Navy)
Small four-stage solid rocket used at NOTS in attempt to orbit NOTSNIK (in 1958) from aircraft at 35,000 ft. See also CALEB. Allegheny Ballistics Lab.

PIMA (Army)
Booster rocket developed for launching the Signal Corps SD-2XAE-1 surveillance drone. Operational. Gabriel—Rocket Power.

PINCUSHION (ARPA)
Radar system for ZEUS tests in the Pacific. A five-story high, multiple-beam radar, capable of detecting lethal missiles among decoys by analysis of disturbed space. Part of Project DEFENDER experimental studies in ballistic missile tracking. 1961 delivery date. Raytheon.

PINPOINT (Army)
Guidance system for ballistic missiles. Refinement of the ATRAN guidance system. Goodyear.

PIONEER I (NASA)
First of a series of space-probe missions, including lunar and deep-space probes. PIONEER is to be differentiated from the EXPLORER series which are earth satellites. PIONEER series began in 1958 with PIONEER I which was launched 11 October 1958 but did not reach escape velocity. First apogee, 71,300 miles. Re-entered over South Pacific 43 hr, 17.5 min after take-off. Was the first outer-space shot. PIONEER launchings made by Army JUPITER C and JUNO II and Air Force THOR-ABLE and ATLAS-ABLE.

PIONEER II
Rocket lofted 8 November 1958 but aborted because of an ignition failure of the third stage. Re-entered atmosphere 42.4 min after take-off.

PIONEER III
Rocket lofted 8 December 1958. Achieved speed slightly below escape velocity. First apogee, 66,654 miles, re-entered atmosphere over French Equatorial Africa 38 hr, 6 min after take-off. PIONEER III discovered the outer Van Allen belt.

PIONEER IV
U. S. deep-space probe lofted into an escape-earth trajectory in the direction of the moon. Passed within 35,000 miles of the moon; now in solar orbit. Launched by JUNO II booster 3 March 1959. Length, 20 in. Diameter, 9 in. Weight, 13.4 lb. Perihelion, 0.9871 AU; aphelion, 1.1421 AU. Period, 397.75 days around sun.

PIONEER V
U.S. space probe lofted 11 March 1960 by THOR-ABLE booster. Weight, 94.8 lb. In solar orbit between earth and Venus orbits. Perihelion, 0.8061 AU; aphelion, 0.9951 AU. Period, 311.64 days around the sun. Purpose: to test long-range radio communications; radiation, space environment measurements. Transmitted data for more than 20 million miles before fading out.

PLASMA PROBE
See EXPLORER X.

PLASTINAUTS
Term applied to three dummy astronauts to be used by Air Force in experiments to explore the possible effects of exposure to radiation in space on the human organism. The plastic surface of the dummies can absorb radiation in a way similar to human tissue.

PLATO (Army)
Transportable guided antimissile system. Program begun in 1955 with competitive studies by Cornell Aerolab and Sylvania Electric. Cornell project canceled as not warranted because of development of NIKE-ZEUS.

PLOVER (Navy)
Small-sized drone with longer range than GORGON from which it was developed. Ramjet engine. Radio-controlled. PLOVER contract to Martin in effect from June 1949 to May 1953.

PLOWSHARE
Nuclear testing discover peaceful uses of the atom with emphasis on those of predominantly economic importance. May be resumed.

PLUTO
Nuclear ramjet engine, forerunner of supersonic, low-altitude
missile SLAM. Used to demonstrate feasibility of high-tempera-
ture air-cooled reactors for ramjets. Study program included
simulated flight on a nuclear reactor (TROY 2A) built for this
study. Estimates of total cost of PLUTO through the flight-test
stage are as high as $500 million, with Air Force providing
75% of the cost. DOD and Air Force have already spent $100
million on PLUTO. Tests to be conducted in Nevada. U.S.
Lawrence Radiation Lab./Marquardt for studies.

PM-1
Packaged power plant developed by Martin.

PMR
Pacific Missile Range.

POBEDA
See T-1.

POGO
Operation at White Sands Proving Ground, to provide high-
altitude radar targets. New Mexico College of Agriculture and
Mechanical Arts.

POGO (NASA)
Polar-Orbiting Geophysical Observatory. Designation, S-50.
Part of OGO program. To be used mainly for studying atmosphere and ionosphere between 170 and 650 miles. Same dimensions as EGO. Scheduled to be launched from Pacific Missile
Range by THOR-AGENA B boosters. First shot set for early
1964.

POGO-HI II (Army)
Training target missile system. Military designation, E-3C.
Drone used for training. Aeronca.

POL-1; POL-2 (U.S.S.R. - Air Force)
Two-stage target missile designed to gather meteorological
data and investigate thermal heating for IGY studies. Recoverable by means of nose cone parachute. POL 1: Length, 14.27 ft.
Weight, 2400 lb. Payload, 9.2 lb. Programed guidance. Solid
propellant. Upper stage reported to be M-100A. POL 2: Length,
about 25 ft. Weight, 25,950 lb. Solid propellant. Payload weight,
90 lb. Inertial programed guidance.

POLARIS (Navy)
Submarine- or surface ship-launched IRBM. Surface-to-surface.
Length, 47 ft. Weight, 28,000 lb. Inertial, celestial guidance.
Range, 1500 miles. Operational. SUPER POLARIS (2500-mile
range) under R&D. Engine: X-250A2; XM33. Lockheed, prime;
G. E./M. I. T., guidance and fire control; Aerojet-General,
propulsion; Lockheed, re-entry vehicle.

POLECAT (Army)
Missile capable of being fired from a recoilless rifle. Guidance
study contract to Bulova Research.

POLKADOT BALLOON
See EXPLORER IX.

POLLUX
See PERCHERON.

POMS
Panel On Meteorological Satellites. Headed by NASA repre-
sentative. NASA chief of meteorological satellites also on POMS.

PONTUS (ARPA)
Research program in U.S. universities on superstrength, radia-
tion-resistant, high-temperature-service materials.

PORPOISE
Underseas oceanographic vehicle to be used in sonar research
by ONR-Navy Hydro Office. Length, 12 ft. Diameter, 21 in.
Ling-Temco Vought.

POUNDER
May be another term for DEFENDER, or separate project.

PRESS (ARPA)
Pacific Range Electromagnetic Signature Study. Part of Project
DEFENDER. Radar systems of advanced design and other
sensing devices to be installed on Roi-Namur Island. Connected
with exo-atmosphere and terminal physics of ballistic missile
flight. A more sophisticated version of DAMP study. Army,
WECO.

PRINCIPIA (ARPA)
Study project on solid propellants. Purpose: development of
propellant with I_{sp} higher by 10 to 20% than those of propellants

now under development. Areas investigated include thermo-chemistry and thermodynamics, synthetic chemistry, and formulation and evaluation of propellants.

PRISM (Navy)

PRogramed Integrated System Maintenance. A system by which constant readiness of the Navy's weapon systems is ensured. First used on the MARK 56 gunfire control system. Because of PRISM success it has been designed to fit the MARK 68 system. This technique is not related to the reliability concept also called PRISM.

PRIVATE A (Army)

Experimental missile used as a test vehicle for a CORPORAL type of guided missile. First of a series of rockets developed by JPL. Main purpose: to provide experimental data on the effect of sustained thrust on a fixed-fin stabilized rocket. Also to learn how to use booster rockets to launch missiles. Length, 8 ft. Weight, loaded, about 500 lb. Range, about 11 miles. Solid propellant. Twenty-four vehicles fired December 1944.

PRIVATE F (Army)

Surface-to-surface missile, similar to PRIVATE A. Major purpose: to test effect of a lifting surface on a guided missile. Seventeen PRIVATE F's fired in April 1945. Tests proved that a missile with lifting surface required flight-control equipment for normal flights.

PROFAC

PROpulsive Fluid ACcumulator. Ramjet-propelled vehicle to orbit earth at 60- to 70-mile altitudes for fueling outbound space vehicles. Proposed by Northrop in 1960.

PROJECT 3059 (Air Force/NASA)

Program to develop a new generation of million-lb-thrust class of solid boosters. In research to determine feasibility. Study contracts to United Technology, Aerojet-General, Thiokol, and Lockheed Propulsion (Grand Central Rocket). Total amount spent on Project 3059, approximately $15 million. Air Force may award $12 million more for follow-on contracts.

PROJECT 7969 (Air Force)

Project to put man in space, 1956.

PROJECTMAN (Army)

Provision of assortment of weapons to deter or fight a total nuclear war, or local aggression with nonnuclear weapons.

PROP

Planetary Rocket Ocean Platform. Buoyant sea base for launching large chemical or nuclear rockets, or space vehicles. Proposed by Aerojet-General.

PROSPECTOR (NASA)

Third program of lunar investigation, to be made by soft-landing, mobile, instrumented vehicles. Early vehicles, capable of traveling up to 50-mile distances by remote control, to explore best landing sites on the moon for forthcoming manned APOLLO missions (lunar craters vs. mountain slopes vs. bare rocks). Later PROSPECTOR vehicles to be altered to transport heavy loads (e.g., jeeps and radiation shelters) to the moon, and on its surface, in support of manned lunar stations. Instrumentation on the missions of these roving vehicles may include integrated radar-TV; command and control subsystems (for steering, stop-start, maneuvers, speed, manipulator control, TV operation, experiment control); low-power telemetry transmitter (for return to earth of such data as temperatures, pressures, vibration, or other data not presentable by TV); a driving force vs. a penetration depth measuring device; a device to measure density and void fractions of lunar material (proposed by Space General); and other scientific experiments ranging from measurement of physical and electromagnetic properties to tests of effects of lunar environment on biological systems. PROSPECTOR vehicles scheduled for launch in the 1964-1966 time period, SATURN booster. May be canceled. JPL.

PSR-1 (Argentina – Army)

Surface-to-surface antitank missile capable of being launched by one man. Solid propellant. Length, 3.8 ft. Launch weight, 19.5 lb. HE warhead. R&D.

PTV

Propulsion Test Vehicle.

PUFFIN (Navy)

Part of the KINGFISHER program. Military designation, XAUM-N-6. Composed of a torpedo with attached wings, a V-type stabilizer, and a pulsejet propulsion system. Radar-controlled. Weight, 1300 lb. Payload, 500 lb. Range, 10 to 20 miles. Speed, Mach 0.7. Canceled, 1949. McDonnell.

PUFS (Navy)
Proposed Underwater Fire Control Feasibility Study. Anti-submarine project based on advanced, wide-range sonar. Hardware being developed by Electric Boat.

PURR-KEE (Navy)
Sounding rocket designed to loft payloads from 2,000 to 18,000-ft altitudes, using DEACON, NIKE, and DAN boosters. In R&D. American Machinery and Foundry.

PUT-PUT (Air Force)
Scale model, 3 ft in diameter, of the Project ORION vehicle, in which 3-lb HE charges simulate the nuclear explosions. Model weight, 300 lb. PUT-PUT tested feasibility of bomb propulsion. See ORION. General Atomics, prime.

PVO (U.S.S.R.)
Soviet Anti-Air Defense Force. See M-100 (U.S.S.R.).

PYE
Antitank short-range surface-to-surface support weapon. Length, 5 ft. Weight, 80 lb. Solid propellant. Visual command link guidance system. Company-funded development by Pye.

PYE P.V. (Great Britain)
Surface-to-surface antitank missile. Solid propellant. Proposal by Pye.

PYE WACKET
Air defense missile designed to be fired from tail of a bomber. Purpose: to destroy interceptors at longer ranges than conventional automatic weapons. R&D. In-house development at G.D.-Astronautics.

PYTHON (Great Britain)
Surface-to-surface antitank missile. Solid propellant. Length, 5 ft. Weight, 80 lb. Wire guidance. In R&D. Proposals submitted. Pye.

Q

Q-2, Q-2A, Q-2C (Air Force)
Military designations for aerial target drone FIREBEE. Ryan.

Q-4B (Air Force)
> Supersonic target drone. Speed, Mach 2. Canceled by Air Force, but prototype and test missiles remaining were transferred to Navy. To be used as targets for TALOS surface-to-air missiles. Radioplane.

Q-5
> See KINGFISHER.

Q-12 (Air Force)
> Target drone launched from NIKE-AJAX ramps or released from aircraft. Speed, Mach 2. Liquid-propellant rocket engine. Same as Navy XKD2B-1. Operational, 1962. See KATYBIRD KX2B-1. Beech.

Q-20
> Target drone. Flight control by Lear-Astronics.

QB-47 (Air Force)
> B-47 aircraft converted to a target drone. Lockheed.

Q-BALL
> Air data sensor for use on the X-15. With modification, to be used to measure angles of attack on SATURN. Developed by Northrop.

QDRI (Army)
> Qualitative Development Requirements Information. Program in which exchange of information on ordnance requirements provides a means of informing industry of Army needs. Forms basis for proposals from industry. Initiated in 1958-59.

QF-80 (Air Force)
> Target drone with remote guidance and control. Ryan.

QF-104 (Air Force)
> Target drone developed from the F-104 plane. Lockheed.

QUAIL (Air Force)
> Formerly Green Quail. Air-to-surface diversionary weapon. Decoy. Gyro autopilot guidance. Range, more than 200 nm. Ceiling, more than 50,000 ft. Deployed at SAC air bases. Operational in 1960. Designed to serve as a decoy for a strategic bomber and also as a vehicle to carry ECM equipment. McDonnell, prime; Sumners Gyro, guidance; G.E., propulsion; Thompson Ramo Wooldridge, ECM equipment.

QUASER
Quantum Amplification by Stimulated Emission of Radiation. Acronym proposed for general classification of MASERS, LASERS, and IRASERS, which vary only in the region of frequency in which they operate.

QUICKGLOBE (Air Force)
Program for automatic data processing and display system to present data for quick decisions during attack. IBM analysis and design study.

QX-10 (Air Force)
Target drone in R&D.

R

R-20 (France)
Reconnaissance drone developed from the CT.20 jet target drone. Range, 90 miles. In flight-test stage. See also MM.20; SM.20. Nord Aviation.

R-422B2 (France - Army)
Surface-to-air, solid-fueled missile. Length, 30.5 ft. Weight, 3500 lb. Radar-command and homing guidance. Range, about 60 nm. Speed, Mach 2 plus. Designed to intercept high-altitude supersonic bombers. Matra.

R-511 (France - Air Force)
Air-to-air missile carried aboard French interceptors. Range, about 5 miles. Supersonic speed. Radar-homing guidance. To be replaced by advanced R-530 missiles in 1962. Operational. Matra.

R-530 (France - Air Force)
Air-to-air missile in advanced development and testing phase. Either semipassive automatic homing head or IR guidance. Speed, about Mach 2. Also offered to NATO. Matra.

RA-1,2,3,4,5
See RANGER.

RAC
Research Analysis Corp. Nonprofit company succeeding Johns Hopkins' Operations Research Office (ORO). Expected to take over large number of ORO personnel and use ORO's head-

quarters near Washington. To complete ORO's unfinished studies. RAC studies to cover basic research, military economics and costing, weapons systems evaluation, strategic and management systems, and information and controls systems.

RACEP
Random Access and Correlation for Extended Performance. Radio-telephone system, 15-mile range, for ground-to-ground communications. Permits a great number of simultaneous conversations. System also applicable to air-to-air and air-to-ground communications. Air Force has bought six units. Martin.

RADIATION LABORATORY
See EXPLORER VIII.

RADIATION SATELLITE
See EXPLORER VII.

RADOP
RADar/OPtical. Towed target systems for air-to-air and surface-to-air missile training. Targets made of light, bomb-like-shaped plastic, with radar reflectors. Various models: DF-4R, subsonic; DF-4RC; DF-4MFC, infrared homing heads; DF-6MFC, supersonic. Del Mar Engineering Labs.

RAF
Royal Air Force.

RAGMOP (Air Force)
Classified communications project. Budgeted for $1.7 million, FY 1961.

RAILS (Army)
Remote Area Instrument Landing System. For helicopter IR capability.

RAM (NASA)
Three-stage booster similar to SCOUT. Solid propellant. First stage an Aerojet Jr. motor with 57,000-lb thrust. Second and third stages to be ANTARES and ALTAIR. Booster considered by NASA.

RAM (NASA)
Radiation Attenuation Measurement rocket. Weight, 7 tons.

RAM (Navy)
Research Aviation Medicine. Basic research and development into aerospace medical techniques, particularly bioinstrumentation for manned space flight. Naval Medical Research Institute.

RAM (Navy)
Antitank missile fired by F4U Corsairs in the Korean War. Discontinued.

RAMAC
An IBM computer.

RAMBLER (NASA)
High-powered, high-altitude sounding vehicle. Capable of launching satellites into orbit. Being developed by team composed of Curtiss-Wright, Vickers, Lear, Wyandotte Chemical, and Spartan Electronics, and directed by University of Michigan.

RAMOS (Navy)
Part of ATLANTIS program.

RANDOM BARRAGE
See RBS.

RANGEMASTER (Army)
Ground-launched aerial reconnaissance drone. Turboprop. Length, approximately 15 ft. Weight, loaded, 1150 lb. Altitude, about 9 miles. Speed, 400 plus mph. RANGEMASTER air-launches the RP-76A, a rocket-powered target drone, then returns to base under radio control while target drone flies on. Both drones recoverable by parachute. Radioplane.

RANGER (NASA)
Number of rough-landing lunar probes, nine presently scheduled. Designations, RA-1 to RA-9. RANGER 1, 2 planned to test basic spacecraft technology for lunar and other missions. This includes a celestial attitude stabilization system, a high-gain directional antenna, an advanced communication system, solar cells, and other components. Also, experiments are conducted to investigate electromagnetic and particle radiation and to measure micrometeoroids. RANGER 1: Placed in near-earth orbit (apogee, 312 miles; perigee, 105 miles) rather than highly elliptical orbit originally planned. Lifetime, about 7 days. Weight, 675 lb. Diameter, 5 ft at base of hexagon. Length, 11 ft. Instrumentation worked satisfactorily. ATLAS-AGENA B boost-

er. RANGER 2 attempt on 18 November 1961 also failed when AGENA B did not restart after coast period. No backup for first two RANGERs. RANGERs 3 to 5: Second phase of program to start in early 1962 with aim to place a seismometer and transmitter on lunar surface. Probes to investigate cislunar magnetic fields and charged particles; also make hydrogen and dust measurements. RANGER 5 may be switched from a hard-lander to lunar orbiter for transmitting lunar photographs. RANGERs 6 to 9: Third phase of program. Mission: to send back to earth high-resolution TV pictures of the lunar surface up to moment of lunar impact. Total cost for program estimated at about $140 million, plus R&D. Unit cost, about $10 million. Launch cost, about $17 million. JPL, prime; Aeronutronic, capsule; Hercules, retrorocket.

RAPID
Rocketdyne Automatic Processing of Integrated Data. System which automatically processes data produced by engine tests. Rocketdyne.

RAPIER (Air Force)
Aircraft (F-108). Canceled. NAA.

RASCAL (Air Force)
Air-to-surface strategic, tactical missile. Radio-command guidance. Liquid propellant. Length, 32 ft. Range, over 100 miles. Speed, Mach 2+. Thrust, 6000 lb. Program initiated in April 1946; canceled in November 1958 after expenditure of $448 million. Bell Aircraft.

RAT (Navy-NOTS)
Rocket Assisted Torpedo. Acoustical homing. Solid propellant. Length, 16 ft. Weight, 480 lb. Range, several miles. Operational since 1958. Canceled after expenditure of $15 million. Superseded by ASROC. NOTS.

RAVEN (Navy)
Proposed air-to-surface missile. Range, 500 miles. Under study by Navy. No contract announced.

RAZON (Air Force)
Unpowered guided bomb, similar to AZON, but controlled both in azimuth and range by an additional set of fins. Discontinued. See AZON, TARZON.

RBS (ARPA/Air Force)
Random Barrage System. Part of DEFENDER. Proposed 20 to 100 thousand armed satellites in random orbits to intercept and destroy enemy satellites or missiles. Studies by Thompson Ramo Wooldridge, and STL.

R&D
Research and Development.

RDT&E
Research, Development, Test, and Evaluation.

REAP (NASA)
Rocket Engine Advancement Program. Research for an engine in the million-pound-thrust class begun in 1953. Rocketdyne.

REBOUND (NASA)
Passive communications satellite system initiated in 1961. Calls for two launches and spare. Weight of payload, about 500 lb. 140-ft-diameter satellites to be launched in groups of three, six, or more from simple satellites and spaced in orbit suitable for communications coverage. Circular orbit 1800 to 2500 miles. First launch scheduled for mid-1963 using two ATLAS-AGENAs to place six satellites in orbit. Douglas has contract for orbital placement and engine design specifications. ECHO, REBOUND, and RELAY projects budgeted for $91 million, FY 1962.

RECON (Army)
Surface-to-surface missile in R&D.

RECRUIT
Solid-propellant sustainer engine used with re-entry vehicles. Thrust, 33,800 lb. Military designations: XM-21, TE-29. Used also in MERCURY project. Thiokol.

REDEYE (Army)
Surface-to-air shoulder-launched antiaircraft missile. Designed to intercept low-flying jets. In R&D. To be operational in 1962. Marines to use REDEYE also. Rumored canceled. Convair, prime; Philco/Convair, Atlantic Research, propulsion.

REDHEAD/ROADRUNNER (Army)
Target missile. Military designation, NA-273. Length, 19 ft. Diameter, 1 ft. Speed, Mach 2. Altitude to 50,000 plus ft. Ramjet engine. Parachute recovery. North American.

REDSTONE (Army/NASA)
Surface-to-surface vehicle. Liquid propellant. Inertial guidance. Length, 63 ft. Weight, 40,000 lb. Range, 200 miles. Speed, Mach 5. Thrust, 75,000 lb. With upper staging, becomes JUNO I. Phased out 1961. Largely replaced by PERSHING. Chrysler, prime; Ford Instrument, guidance; Rocketdyne, propulsion.

RED TOP (Great Britain - RAF)
Air-to-air missile. Solid propellant. IR guidance. Range, over 9 miles. Speed, Mach 3. 68-lb HE warhead. Successor to FIRE-STREAK on various aircraft. Same concept as U.S.N. EAGLE. De Havilland Propellers.

REDWING
See ASP (Navy).

REGAL
Intricate guidance system under development to govern range, guidance, approach, and landing of modern high-speed airplanes. Operational for New York City about 1963. To be installed later in Chicago, Los Angeles, Washington, and Miami.

REGULUS I (Navy)
Navy version of the Air Force MATADOR. Surface-to-surface missile. Military designation, SSM-N-8A. Length, 30 ft. Weight, 14,500 lb. Command guidance. Solid propellant. Turbojet-powered. Range, 500 to 600 miles. Thrust, 4600 lb. Speed, over 700 mph. Carried aboard submarines, surface ships. Canceled. Chance-Vought.

REGULUS II (Navy)
Surface-to-surface tactical missile. Military designation, SSM-N-9A. Length, 55 ft. Weight, 35,000 lb. Self-homing; command guidance. Solid propellant. Turbojet-powered. Range, over 1000 miles. Thrust, 17,000 lb. Speed, Mach 2. Nuclear warhead. Was used also as target drone. Program canceled December 1958 on grounds that the missile had become obsolescent, after expenditure of $90 million. See MEGABOOM, ZEL. Chance-Vought, prime and guidance; Aerojet-General, propulsion.

RELAY (NASA)
Program of active, low-altitude, real-time communication satellites. Four shots scheduled. First two satellites to be lofted by THOR-DELTA into orbit with 900-mile perigee, 3000-mile apogee. Weight of satellites, 100 lb. First launching sched-

uled for last half of 1962. Purpose: to transmit wideband signals, TV signals, multichannel telegraphy, data-handling between east U.S.A. coast and Western Europe. Cost: $450,000 to $600,000 for each satellite, plus cost of booster and launch. With ECHO and REBOUND programs, budgeted for $91 million, FY 1962. RCA, prime. RELAY II: New program under consideration for satellite in 6-hr subsynchronous orbit, either circular or elliptical, at inclinations of 60 or 63.4°. ATLAS-AGENA B booster. NASA may let contract early 1962.

RELLA (Navy)
Air-launched probe used to carry sensing packages for the measurement of atmospheric conditions. Under R&D.

RENAE (Navy)
Ocean weather satellite to observe weather conditions in remote waters for coordination with the fleet tactical data systems. RCA study.

REPPAC III
REpetitively Pulsed Plasma ACcelerator propulsion engine for use in space. Reportedly, 7 kw of power needed to operate the engine at 32% efficiency. G.E.–Space Science Laboratory.

RETORC (Navy)
REsearch TORpedo Configuration. A research program to develop advanced torpedo design RETORC I: development of small torpedoes weighing less than 500 lb. NOTS, Pasadena, Calif. RETORC II: development of torpedoes weighing up to 4000 lb. Projects divided between Pennsylvania State University, Ordnance Research Laboratory, and Naval Underwater Ordnance Station, Newport, R.I.

RFP
Request For Approval.

RHEINBOTE (Germany)
Four-stage, high-acceleration missile designed for long-range, surface-to-surface operation. Fin-stabilized. Weight, loaded, 3773 lb. Length, about 37 ft. Speed, 3700 mph. Range, 100 to 135 miles. Some 60 missiles fired against Antwerp, but with little effect, beginning November 1944. See T-5 (U.S.S.R.).

RIFT (AEC/NASA)
Reactor In Flight Test. A test program which calls for flight testing the nuclear rocket NERVA engine that may be used in

SATURN upper stages in 1966-67. A 350-mile orbit planned for tests. The program might include development of space operation techniques such as rendezvous, docking, and booster recovery. A contract to conduct tests expected to be awarded in 1962. Study contracts awarded to Lockheed, Martin, Douglas, and G. D.–Astronautics. Later Douglas was dropped from the competition.

RIGEL (Navy)

Large tactical missile for shore bombardment. Military designation, XSSM-N-6. Submarine-launched. Length, 46 ft. Range, 550 miles. Later configurations were: 26 ft long approximately; weight, 6200 lb at launch; and reduced range of slightly over 100 miles. Program conducted 1946 to 1953. Grumman.

RISE (Air Force)

Research In Supersonic Environment. Military designation, XSM-64. Target drone program using the X-10.

RISP

Recoverable Interplanetary Space Probe. Recoverable probe designed to fly to Mars, photograph approximately 40% of the planet, return with recoverable package, and land in the Gulf of Mexico. M.I.T. Instrumentation Lab; Avco; Thiokol.

RITA '

Reusable Interplanetary Transport Approach. A cone-shaped, single-stage, self-contained nuclear rocket spacecraft. Fifteen stories high. Reusable. Includes "bio-well" as a retreat during solar flares or other radiation hazards. Might have artificial gravity for crew sections. RITA A model could lift 10,000 lb to the moon. RITA B could lift 25,000 lb to the moon. With space station refueling, payloads could be raised to 38,000 lb for A, and 90,000 lb for B. SATURN booster first stage. Conceptual. Douglas.

RL-10 (NASA)

Liquid hydrogen rocket engine. Formerly designated LR-115. Slated to power the CENTAUR. Thrust, 15,000 lb. Has 30% greater efficiency than current rocket engines using conventional chemical fuels. Developed by Pratt & Whitney.

RM-1100

Engine used for upper stage of ASP 1, ASPAN. Liquid propellant. Thrust, 5982 lb. Marquardt–Cooper.

RM-1400
Booster used for ASP 1, ASPAN. Liquid propellant. Thrust, 2850 lb. Marquardt—Cooper.

RM-2110
Liquid-propellant booster for ROKSONDE·100. Thrust, 2950 lb. Marquardt—Cooper.

RM-2210
Liquid-propellant booster used for ROKSONDE 200. Thrust, 1900 lb. Marquardt— Cooper.

ROBIN (Air Force/Navy)
ROcket Balloon INstrument. Sounding rocket consisting of ARCAS and LOKI-DART rockets carrying a deflated weather balloon. Developed by Air Force Cambridge Research Center and ONR. Experiment continued to November 1960. Atlantic Research/ Schjeldahl.

ROBO
Early project by Boeing (distantly related to DYNA-SOAR) for unmanned glide missiles.

ROBOT 304 (Sweden—Air Force)
Also JAKTROBOT. Designation BO-4. Air-to-surface missile designed for use against naval targets. Designation, BO-4. Solid propellant. Radio command. Range, about 3 miles. Speed, Mach 1. Deployed with the Swedish Air Force. Operational. Swedish Guided Weapons Bureau and Royal Swedish Armed Forces Research Establishment.

ROBOT 315 (Sweden—Navy)
Surface-to-surface tactical missile. Length, 24 ft. Weight, 3000 lb. Solid propellant. Range, 10 to 20 miles. Speed, Mach 0.9. Deployed aboard Swedish destroyers. Torpedo boat model known as Sjorobot. Swedish Guided Missile Bureau and Royal Swedish Armed Forces Research Establishment.

ROBOT 322 (Sweden—Army)
Surface-to-air missile. Twin ramjet engines. Solid propellant. In R&D. Swedish Guided Missile Bureau and Royal Swedish Armed Forces Research Establishment.

ROBOTTI (Italy)
Family of unguided battlefield rockets. Supersonic speed. Moto-fides.

ROC-1 (Air Force)
Also ROC 00-1000. Military designation, VB-10. Rocket developed in 1941. Capable of carrying a 1000-lb payload. Originally a radar beam rider. Also equipped with electric eye to home on light. Project dropped at end of World War II. Douglas.

ROCKAIR (Navy)
FFAR rocket instrumented for sounding experiments. Length, 4 ft. Weight, loaded, 18.5 lb. Payload weight, 2.5 lb. Altitude, 35 miles. Speed, 2500 mph. University of Maryland Physics Department designed the first payload packages. Program initiated and completed in 1955.

ROCKAIRE (Air Force)
Solid-propellant test rocket capable of boosting a 40-lb payload. Actually a DEACON rocket launched from aircraft. Length, nearly 9 ft. Weight, loaded, about 180 lb. Capable of reaching 28-mile altitude. Speed, 2900 mph. Purpose: to gather atmospheric data to about 40 miles, but this was not achieved. Four firings made in December 1956 from F-86D airplane which had climbed to nearly 7-mile altitude before firing the ROCKAIREs. Douglas.

ROCKET—140MM (U.S.S.R.—Army)
Battlefield rocket with four pop-out folding fins. Solid propellant. Length, over 3 ft. Weight, 120 lb. Carries 55-lb warhead. Unguided. Range, 9900 yd. Speed, Mach 1.5. Seen first in November 1959 in Moscow. Mounted on truck. Used by U.S.S.R. tank forces.

ROCKET—200MM (U.S.S.R.—Army)
Battlefield rocket mounted on trucks. Solid propellant. Length, 10.5 ft. Weight, over 425 lb. Can carry 125-lb HE warhead. Unguided. Range, 20,000 yd. Speed, over Mach 2. Reportedly 75 launchers in each Soviet tank army. First seen in 1954. Operational status.

ROCKET—240MM (U.S.S.R.—Army)
Solid-propellant battlefield rocket. Length, 4.2 ft. Weight, 248 lb. Angled venturis. Unguided. Range, 7700 yd at Mach 1.5 speed. Capable of carrying a 100-lb HE warhead. Mounted on truck or tractor. Reportedly 12 ROCKET-240MMs in each Soviet tank division. Seen first in 1953.

ROCKET—280MM (U.S.S.R.—Army)
Solid-propellant battlefield rocket. Length, 17 ft. Weight, about 1000 lb. Can carry 475-lb HE warhead. Unguided. Range, 23,500

yd. Speed, more than Mach 1. Mounted on heavy trucks. Considered to be standard on all Soviet artillery regiments. Seen first in 1957.

ROCKETSONDE (Air Force/ARCD)
Rocket fired from a launch tube suspended from three parachutes at high altitude. Purpose: to obtain high-altitude weather information over ocean and polar regions. Developed by Convair.

ROCKOON
DEACON rockets launched from balloons at about 80,000-ft altitudes and reaching altitudes of about 340,000 ft over Thule, Greenland, and the Pacific Ocean. See DEACON.

ROCMIZ
Multistage sounding rocket whose performance is similar to NIKE-CAJUN. Solid propellant. Completed prototype built by Zimney.

ROCOPTER
A manned or unmanned space vehicle which combines a booster with a capsule provided with rotary wings for re-entry. Blades, rotating with the wind, open gradually as the vehicle pursues a ballistic trajectory to earth. Purpose: to supply troops or to transport mail, freight, or passengers. Patented in 1961 by a Huntsville scientist.

ROK (Korea)
Rocket with 50-mile range.

ROKSONDE 100, 200
Research vehicle capable of lofting small payloads to 200,000-ft altitudes. Engines: RM-2110; RM-2210. Operational. Marquardt—Cooper.

ROSE
Rising Observational Sounding Equipment. Mylar balloon designed to provide operational weather data at 0- to 100,000-ft altitudes. Free-rising; super-pressure. Tested successfully by Air Force. Balloon composed of 12 panels colored alternately red and silver in order to facilitate visual determination of its rotation. Schjeldahl.

ROTC
Reserve Officers Training Corps.

ROTOCHUTE (Air Force)
Concept of returning capsules back to earth by whirling them like leaves, with rotary blades. In R&D. Kaman Aircraft.

ROVER (AEC/NASA)
Nuclear propulsion program conducted by Los Alamos Scientific Laboratory and Atomics International. Incorporates KIWI and DUMBO projects to develop a superthrust engine capable of lifting 1000-lb payloads to 1000-mile altitudes. KIWI A reactor power runs conducted in 1959 and 1960. KIWI B reactor power runs to be made through 1962. Reactor-in-flight (RIFT) test program to launch first NERVA engine in 1966-67 time period.

RP-70 (Army)
Target drone with solid-propellant rocket engine. Length, 9.5 ft. Weight, loaded, 305 lb. Speed, Mach 0.95. Altitude, more than 11 miles. Made principally of plastic; expendable. Unguided; controlled entirely by an autopilot. Similar to Navy XKD4R-1. Radioplane.

RP-71, 71F
See SD-1.

RP-76A (Army)
Surface-to-air target. Either ground- or air-launched. Solid-propellant rocket motor; radio-controlled. Length, over 9 ft. Weight, loaded, 300 lb. Speed, Mach 0.9. Altitude, more than 11 miles. May be launched either from a F-89 Scorpion or the RP-77D vehicle, also a target drone. See also RANGEMASTER. Radioplane.

RP-77D
See RANGEMASTER.

RP-87, RP-88
Advanced versions of the target drone OQ-19.

RRS
Reaction Research Society. National amateur rocket group. Hold rocket mail flights on test sites established by the organization.

RS-82; RS-132; RS-132A (U.S.S.R.—Air Force)
Air-to-air missile used for training. Aircraft version of the GVAI.

RSA-54 (Switzerland)
Earlier model of RSD-58, liquid-fueled missile for use by combat units. See RSD-58.

RSC-57 (Switzerland)
Training model of a surface-to-air liquid-fueled missile. For use by combat units. See RSD-58.

RSD-58 (Switzerland)
Surface-to-air liquid-fueled missile deployed with combat units in Switzerland, Italy, and Japan. Range, more than 18 miles. Speed, Mach 2.5. Ceiling, 65,000 ft. Operational. See RSA-54 and RSC-57. Contraves AG; Oerlikon.

RTV
Research Test Vehicle.

RTV-A-1, 1a, 1b
See AEROBEE-HI.

RTV-A-3
See NATIV.

RTV-N-6
See TALOS.

RTV-N-8al
Also RTV-N-10; RTV-N-10b. Military designations for Navy models of AEROBEE-HI.

RTV-N-12
See VIKING.

RTV-1
Research test vehicle using beam-rider guidance. Time period, 1954. Royal Aircraft Estab., England.

RUBY (Air Force)
Study of the factors affecting radar return data received from an object in space. Cornell Aero. Lab.

RUN (Navy/ONR)
Remote control ocean bottom crawler. Purpose: to transport, assemble, and install instrumentation assemblies on the ocean floor. Can reach depth of 20,000 ft. Contracts may go to Hughes, Orbitran, and General Mills.

RUSPANO
Russian-Spanish dictionary compiled by machine translation.
Universidad Nacional Autónoma de México.

RV-A-10 (Army)
Surface-to-surface test vehicle in HERMES A-2 program. G.E.

RVX-2
See ARC.

S

S-3 (NASA)
Series of four satellites to study energetic particles in the Van
Allen belt and sun; cosmic rays. Weight, 82 lb. Perigee, 150
miles; apogee, 40,000 miles. THOR-DELTA (6) booster. First
launch 15 August 1961.

S-3 (NASA)
Second-stage engine for SATURN C-2 version. 200,000-lb
thrust. See SATURN C-2; J-2.

S-4 (NASA)
Second-stage engine for SATURN C-1. 20,000-lb thrust.

S-6 (NASA)
Satellite to study upper atmospheric structure. Scheduled to be
launched in early 1962 by THOR-DELTA. Apogee, 650 miles;
perigee, 125 miles. Weight, 370 lb.

S-15 (NASA)
See EXPLORER XI.

S-16, S-16A (NASA)
See OSO.

S-17 (NASA)
See OSO.

S-26
See ANNA.

S-27
See TOPSIDE SOUNDER.

S-45 (NASA)
Satellite to measure properties of the ionosphere. Weight, 75 lb.
JUNO II booster. Failed February 1961. Backup satellite failed
27 May 1961.

S-48 (NASA)
Six-Frequency Topside Sounder satellite to measure electron
density of ionosphere approximately 200 miles above earth.
SCOUT booster. Experiments to include telemetry, three dual-
channel transceivers, solar batteries, 4000 exterior solar cells,
a tracking beacon, and sounding antennas. Satellite designed
and built by Cutler-Hammer and National Bureau of Standards.

S-49
See EGO.

S-50
See POGO.

S-51
See U.K. I.

S-55 (NASA)
Micrometeoroid-measuring satellite. Weight, 115 lb. SCOUT
booster. Launch 1961.

S-56
See BABY ECHO.

S-60-5A
See TIROS I.

S-60-6A; S-60-10
See TRANSIT I.

S-60-11
See DISCOVERER XII.

S-60-12
See DISCOVERER XIII.

S-60-13
See ECHO.

S-60-14
See DISCOVERER XIV.

S-60-15
COURIER 1-A. Launched 18 August 1960. THOR-ABLESTAR booster. Failed.

S-60-16
See DISCOVERER XV.

S-60-17
COURIER 1-B. Launched 4 October 1960. THOR-ABLESTAR booster. Successful.

S-60-19
See DISCOVERER XVI.

S-60-21
See DISCOVERER XVII.

S-60-22
TIROS II. Launched 23 November 1960. THOR-DELTA booster.

S-64 (NASA)
Second in series of four satellites to study radiation and energetic particles in Van Allen belt and sun; solar wind; the interplanetary magnetic fields; and the distant regions of earth's magnetic field. (First satellite in series was S-3.) S-64 scheduled to carry equipment to measure the particle plumes, types and energies in the outer edge of the outer Van Allen radiation belt as functions of position, direction, and time. To carry several particle-detector systems, optical aspect sensor, and magnetometer for aspect and field measurement. Also to study long-period variations in the primary cosmic-ray intensity. Scheduled to be placed in 24-hr orbit at 22,300-mile altitude.

S-438-L (Air Force)
Intelligence data-handling system.

SABER
Solid-propellant rocket which is an improved VIPER. Length, 8 ft. Weight, 210 lb. Designed for sounding, drone-boosting, and sled propulsion. Lockheed Propulsion (formerly Grand Central Rocket).

SAC
> Air Force Strategic Air Command Hq., Offutt AFB, Neb. See also STRIKE COMMAND, a unified TAC-STRAC command.

SACEUR
> Supreme Allied Commander EURope. Responsible to NATO for the defense of the land mass of western Europe.

SACLANT
> Supreme Allied Command atLANTic. Responsible to NATO for the defense of the area north of the Tropic of Cancer to the North Pole. Includes an Antisubmarine Warfare Research Center, La Spezia, Italy.

SAGE (Air Force)
> North American air defense system. Military designations: SS-416L; WS-216L. Burroughs/Systems Development Corp.

SAINT (Air Force)
> SAtellite INspection sysTem. In early phase, acronym for SAtellite INTerception. At one time, called HAWKEYE. Passive antisatellite system to develop rendezvous capability with unknown satellite for inspection purposes. A "coorbiting" satellite rather than an antisatellite satellite. To be orbited by ATLAS-AGENA, possibly by late 1962. Aerospace Corp., technical management. RCA is developing eight payloads. Budgeted for approximately $6 million, FY 1961; $26 million, FY 1962.

SAM
> Surface-to-Air Missile.

SAM-A-7
> See NIKE-AJAX.

SAM-A-18
> See HAWK.

SAM-A-25
> See NIKE-HERCULES.

SAM-N-6
> See TALOS.

SAM-N-7
> See TERRIER.

SAND (NBS)
A routine system for processing information, especially for U.S. Patent Office.

SANE
Scientific Applications of Nuclear Explosions. Meeting of scientists of various disciplines at Los Alamos, N.M., July 1959. Meeting restricted in order to allow free discussion under information security rules.

SAPHIR (France)
See TOPAZE.

SAPO
Computer used for machine translation. Research Institute for Mathematical Machines, Prague.

SARAH (NASA)
Search And Rescue Homing Beacon. Rescue beacon used in the MERCURY project. Formed part of the astronaut's personal equipment. Also placed in the capsule to give accurate location of "ditched" astronaut. Ultra Electronics/Simmonds Precision Products.

SARAH (Navy)
New version of the SIDEWINDER air-to-air missile. Electronic guidance. In R&D. Motorola/Texas Instruments.

SARUS
Small search And Rescue Using Satellites. Payload weight, 20 lb. Proposal to NASA. Space Electronics, Glendale, Calif.

SATAF
Site Activation TAsk Force. An Army Corps of Engineers task force responsible for activating ICBM missile sites (ATLAS, TITAN, MINUTEMAN) for the Air Force.

SATAN (Air Force)
SATellite Active Nullifier. Antisatellite system consisting of satellites carrying missiles capable of destroying ICBM's. Under study. Funded by ARPA.

SATELLAB
Manned space laboratory capable of supporting three scientists and about 600-lb test payload. Proposal in 1960 by Chance-

Vought-Astronautics. Could be placed in orbit for six-year period. SATURN booster.

SATIC

Scientific And Technical Information Center. A national clearing house for scientific and technical information. Advocated by Army Chief of R&D, Lt. Gen. Arthur Trudeau because of the great need to improve communications and aid the exchange of information.

SATIN

SATellite INspector system study. Completed. Designation, SR 79998.

SATRAC

Satellite Automatic Terminal Rendezvous And Coupling. An optical-homing guidance system and coupling device used to rendezvous and couple two satellites. Guidance is for terminal phase and depends on light from the target satellite for optical guidance. Coupling made by hook and ring connection. The rotation resulting from coupling is damped by an attitude control system. Convair.

SATURN (NASA)

Program intended to develop booster capable of sending a manned vehicle around the moon; also, of putting a segmented space station into orbit for assembly there.

SATURN C-1 (NASA)

Three-stage vehicle capable of lofting 19,000-lb payload into a 300-mile orbit, or 5000 lb to the moon. Length, 150 ft. Gross weight, 927,000 lb. Consists of S-1 booster (8 H-1 Rocketdyne LOX-RP engines) of 1.5 million-lb total thrust, plus S-4 (Douglas) second stage composed of 6 Pratt & Whitney RL-10B LOX/H_2 engines, each with 15,000-lb thrust, plus S-5 (G.D.—Astronautics) third stage composed of two Pratt & Whitney RL-10A3 engines, each with 15,000-lb thrust. Scheduled to launch three-man APOLLO A orbital shots in 1965. Also considered for PROSPECTOR missions and first VOYAGER flights. In advanced development. Vehicle funded at $224 million for FY 1962; $200 million contract to Chrysler, December 1961. Appropriations through FY 1962 for construction of facilities for the program are $86.2 million. Total Chrysler prime contract for the S-1 first stage expected to be of the order of $2 billion. First 10 developmental vehicles (being built by Marshall Space Flight

Center) to cost $80 million each. Twenty operational vehicles to be fabricated by Chrysler at Michoud plant.

SATURN C-2 (NASA)

Program, now canceled, for a liquid-propelled booster capable of lofting 45,000 lb into a 300-mile orbit, or 15,000 lb to the moon. All inertial guidance. Was to consist of S-1 booster plus S-2, S-4, and S-5 stages.

SATURN C-3 (NASA)

Program, now canceled, for a liquid-propelled booster capable of putting 100,000-lb payload into low orbit, or 38,000 lb on the moon. Three-stage configuration with first stage composed of 2 F-2 engines, each with 1.5 million-lb thrust; second stage, 4 engines of 200,000-lb thrust each, and third stage, 6 engines with 15,000-lb thrust each. Superseded by SATURN C-5.

SATURN C-4 (NASA)

Program, now canceled, calling for 4 F-1 engines in first stage and capable of lofting 200,000 lb into a 300-mile orbit. Was planned to loft a manned vehicle, e.g., APOLLO, into lunar orbit, or with rendezvous techniques, would have been capable of lunar landings. Superseded by SATURN C-5.

SATURN C-5

Space booster selected for development in lieu of SATURN C-2, C-3, and C-4. To be capable of performing rendezvous-type manned landings on the moon. Can launch 220,000-lb payload into 300-mile orbit or 90,000 lb to the moon. Stage 1 (S-1B or S-1C): 5 Rocketdyne F-1 (LOX-RP); total thrust, 7.5 million lb; Boeing, prime. Contract to Boeing of about $300 million, January 1962. Total cost of S-1B expected to be about $1 billion. For FY 1962, $16 million, and for FY 1963, $92.5 million allocated. Stage 2 (S-2): 5 Rocketdyne J-2 (LOX-H_2); total thrust, 1 million lb, $39 million allocated for FY 1963. Stage 3 (S-4B): one J-2, 200,000-lb-thrust engine. Douglas, prime. Total budget, $250 million. The SATURN C-1 and SATURN C-5 will be capable of fulfilling all APOLLO missions, assuming development of rendezvous techniques. NOVA configuration to be needed for direct lunar assault with manned craft.

SAUCER (Air Force)

Two-year program at Wright-Patterson to investigate the phenomena of "flying saucers." Project abandoned in 1949.

SBAMA (Air Force)
   San Bernardino Air Materiel Area, located at Norton AFB,
   San Bernardino, Calif.

SCANS
   Scheduling and Control by Automated Network Systems. A more
   sophisticated version of PERT developed by Systems Development Corp., Santa Monica, Calif.

SCAR
   SubCaliber Aircraft Rocket. Similar to the 3.5- and 5-in. FFAR's
   used to train rocket-firing airplane pilots in World War II.
   Developed by California Institute of Technology group of the
   National Defense Research Committee, and Navy BuOrd.

SCAR (IGY)
   Special Committee on Antarctic Research. Under International
   Geophysical Year.

SCAR (Navy)
   Submarine Celestial Altitude Recorder. Device with which a
   submarine can take a celestial fix from periscope depth while
   still submerged.

SCHMOO
   Proposed spacecraft design, vase-shaped, powered by a nuclear
   reactor in upper part of the craft. Would include crew and cargo
   modules in lower toroidal section, thus facilitating loading and
   unloading at extraterrestrial bases. Directional control by means
   of four swivel exhaust nozzles in top of vehicle. Diameter,
   about 200 ft. Passengers to be situated near the perimeter.
   Rotation to produce artificial gravity. Douglas.

SCORE (ARPA/Air Force)
   Signal Communications by Orbiting Relay Equipment. Cylin-
   drically shaped satellite launched 18 December 1958 from Cape
   Canaveral. Length, 85 ft. Diameter, 10 ft. Gross weight of payload, 8700 lb; net weight (nose compartment), 168 lb. Equatorial,
   wide elliptical orbit. Inclination to equator, 32.3°. Perigee,
   110.6 miles; apogee, 911 miles. ATLAS booster. SCORE first
   to record and transmit the human voice (President D. D. Eisen-
   hower's) through space. In orbit to 21 January 1959. Superseded
   by COURIER.

SCOUT (Air Force)
> Version of SCOUT booster developed for Air Force. Same suppliers as for NASA SCOUT. See also BLUE SCOUT.

SCOUT (NASA)
> Four-stage "poor man's" solid space booster capable of lofting 150- to 200-lb payload into 300-mile orbit, or 65-lb payload to 8000-mile orbit. Length, 72 ft. Total weight, 36,600 lb. First stage: ALGOL (Aerojet-General). Modified early POLARIS. Weight, 23,600 lb. Thrust, 115,000 lb. Fin- and jet-vane-stabilized. Second stage: CASTOR. An improved Thiokol SERGEANT. Weight, 9300 lb. Thrust, 55,000 lb. H_2O_2-jet stabilized. Third stage: ANTARES. An improved VANGUARD third stage (Hercules Powder). Weight, 2600 lb. Thrust, 13,600 lb. Fourth stage: ALTAIR (Alleghany). Weight, 520 lb. Thrust, 3060 lb. Vought spin-stabilized, similar to VANGUARD third stage. SCOUT used or scheduled for use in MERCURY (for testing capsules), BABY ECHO, EXPLORERs IX and XIII, P-21, P-21A, S-48, S-51, S-55, and TRANSIT. NASA also collaborating with Britain, Canada, and other nations in supplying SCOUTs for scientific research. See also BLUE SCOUT. Cost of development of first nine vehicles, about $17 million. Cost per copy exclusive of R&D, about $750,000, or about $950,000 per operational launch. About $50 million budgeted for SCOUT program in FY 1961. Ling-Temco Vought, prime; Minneapolis-Honeywell, guidance; Aerojet/Hercules/Thiokol/ABL, propulsion.

SCULPTOR (ARPA)
> Study to integrate large payloads of different missions. Outgrowth of ARROWHEAD.

SCWS-649E
> Space Combat Weapon System. Outgrowth of SPAD or RBS.

SD-1 (Army Signal Corps)
> Surveillance drone for propeller-driven aircraft. Military designations: RP-71; RP-71F. Length, 13.4 ft. Weight, loaded, 425 lb. Speed, 180 plus mph. Altitude, approximately 3 miles. Radio-controlled. Liquid propellant. All-weather. Catapult-launched. Recoverable by parachute. Version of OQ-19 target drone. Radioplane.

SD-2
> See FLYING SPY.

SD-3
> See SNOOPER (Army Signal Corps).

SD-4
   See SWALLOW.

SD-5
   See OSPREY.

SDI
   Selective Dissemination of Information. An automatic system
   for matching interests of personnel with available data and
   supplying them with such data. IBM.

SDR-I
   See SNAP II.

SE 4200
   See CAISSEUR.

SE 4400
   See TRIDENT (France).

SEACAT (Great Britain—Navy)
   Surface-to-air short-range missile. Solid propellant. Range,
   about 4 miles. Speed, Mach 1. HE warhead. Deployed aboard
   destroyers and cruisers. Also ordered by Australian, Swedish,
   and New Zealand navies. Initially operational in 1960. Advanced
   version, under development, to have 10- to 11-mile range.
   Short Bros./Harland.

SEAPUP VI
   Two-man submarine. Length, 19 ft. Designed for scientific
   exploration of ocean at depths greater than 1 mile. General
   Mills—Electronics Group.

SEA QUEST
   Oceanographic research vessel operated by Lockheed-California.

SEA SCOUT (Navy)
   Family of four solid-propellant satellite launchers which would
   have mission similar to SCOUT. Would make use of surplus
   POLARIS missile engines as the missiles are supplied with
   improved engines. Partly composed of the first two stages of
   POLARIS plus NASA SCOUT upper stages. Vetoed by Office of
   Defense R&D. Proposed again by Navy.

SEASLUG (Great Britain—Navy)
Surface-to-air medium-range missile. Solid propellant. Length,
19.5 ft. Beam-riding guidance. HE warhead; later to be nuclear.
To be deployed aboard British Navy destroyers. Initially operational in 1960. In production. Armstrong Whitworth.

SECOR (Army)
SEquential COllation of Range satellite. Precision long-range
mapping system. For piggyback launching with Navy TRANSIT.
USACE Map Service.

SECT (Navy)
Program for the development of underwater communications.
Study of means of propelling antennas to the surface from great
depths. Aerojet-General has feasibility study contract.

SEMPER (Marines)
Believed to be an air-launched missile. Reportedly under test
at Pt. Mugu, Calif. No contracts announced.

SEPR
Société d'Etude de la Propulsion par Réaction.

SER
See SNAP II.

SEREB
Société pour l'Etude et la Réalisation d'Engins Balistiques.
An association of missile and aircraft firms in France, estab-
lished in 1959.

SERGEANT (Army)
Surface-to-surface field artillery missile. Solid propellant.
Length, 30 ft. Weight, about 9000 lb. Range, 100 miles. Speed,
supersonic. Inertial guidance. HE or nuclear warhead. Air or
truck transport. Successor to CORPORAL. Engine: XM12.
Military designations: SSM-A-26; FAGMS-S. R&D completed.
Possible cancellation in 1962 because of overlap in range with
other missiles. Sperry, prime and guidance; Thiokol, propulsion.

SERT (NASA)
Space Electric Rocket Test. Program to start in late 1962 to
flight-test feasible engines, using a SCOUT booster. First
flights to be in a ballistic trajectory. Later ones may be orbital
flights. SERT I will test two engines: a cesium ion system

developed by Hughes and a mercury ion engine developed by NASA. Backups will be available. Payload weight for SERT I to be 200 lb. Trajectory will attain 4100-mile altitude and 4000-mile range, so that capsule will be above 200 miles for an hour. SERT II will test an arc jet built by Plasmadyne and an ion engine made by Electro-Optical Systems, Pasadena, Calif.

SES
See SURFACE EFFECT SHIP; GEM; LOTS.

SET (Army)
Sensory Evaluation Test program. To use X-7 drone. Proposed by Lockheed-California.

SETEL
Société Européenne de Téléguidage, consisting of Thomson-Houston, France; Finmeccanica, Italy; Telefunken, Germany; Ateliers de Construction Electrique de Charleroi, Belgium; and Philips, Netherlands.

SHAPE
Supreme Headquarters Allied Powers, Europe. SHAPE Air Defense Technical Center in The Hague, Netherlands. Part of NATO command organization in Europe.

SHARE
An organization of computer users sharing routines, services, and experience in the data-processing field.

SHARK (Navy)
Underwater study made by Hughes in 1954.

SHAVIT SHTAYEM (Israel)
Multistage sounding rocket. Solid propellant. Also known as METEOR II and SHAVIT II. Total weight, 550 to 660 lb. Altitude, 50 miles into the ionosphere. No telemetering equipment carried. Speed, 4900 plus ft/sec. Launched July 1961. Cost exclusive of R&D, about $25,000.

SHEPHERD (ARPA)
Ground-based system designed to detect "dark and silent" satellites in space. Part of ARPA's INSSCC. Detection complexes include Navy SPASUR stations, Army DOPLOC stations, and Air Force computation control center, part of SPACE-

TRACK program. Budgeted $32 million up to FY 1959; $12 million for FY 1960; $18 million for FY 1961. Ford-Aeronutronic.

SHERWOOD (AEC)
Project to produce a "baby" sun with 200 million-degree temperature as a source of energy. Westinghouse Electric contributing.

SHILLELAGH (Army)
Surface-to-surface lightweight missile for close-in support of troops. Nuclear warhead. Two versions: hand-carried and vehicle-mounted. Operational. Ford-Aeronutronic.

SHOT PUT (NASA)
Research vehicle capable of boosting 150-lb payload to a 250-mile altitude. Solid propellant. Operational. Douglas.

SHRIKE (Air Force)
Air-to-surface tactical missile similar to a small RASCAL missile. Length, 22 ft. Weight, 3000 lb. Liquid rocket motor. Range, 30 to 60 miles. Altitude, 6 to 8 miles. Speed, Mach 2. Bell Aircraft.

SHRIKE (Navy/Air Force)
Air-to-surface antiradiation missile system, formerly called ARM. Designed to home in on electromagnetic emission installations. Aircraft-launched. Air Force may use homing system on other missiles. Developed by NOTS. Guidance and control, Texas Instruments.

SHRIMP (Air Force)
Also identified as SICBM, MIDGETMAN, and MINUTEMAN II. Solid-propellant small ICBM project. Weight, 23,000 lb. To be transported and launched from trucks and trains, and have CEP of less than 1 mile. Term also applied to MRBM project. In study phase.

SICBM
Small InterContinental Ballistic Missile. See SHRIMP; MIDGETMAN; JOSHUA.

SIDEWINDER (Navy)
Air-to-air guided missile developed by NOTS. Military designations: AAM-N-7 (Navy); GAR-8 (Air Force); WS-221A. Lightweight. Length, 9 ft. Weight, 155 lb. Range, 6 to 8 miles. Speed,

Mach 2.5. IR homing. HE warhead. Operational. Used successfully in Korean War. SIDEWINDER 1C is second generation.
Has switchable IR and radar-homing warheads. Philco, prime,
under direction of NOTS.

SIGMA (Japan)

Project undertaken by Science Council of Japan for the IGY.
Purpose: to develop a rockoon capable of lofting a 5-lb payload
to an altitude of 55 to 65 miles. Two types of balloons used:
plastic and a cluster of seven balloons. Three types of rockets
used: SIGMA, PI, and WASP. SIGMA and PI designed and developed by Itokawa Lab., University of Tokyo; WASP, by Marquardt—Cooper. SIGMA: Length, 6.9 ft. Weight, 88.5 lb. Payload
weight, 9 lb. Altitude, 43.5 miles. PI: Length, 6.9 ft. Payload
weight, 6.6 lb. Altitude, 62 miles.

SILVER SAINT (Air Force)

Satellite interceptor program. The antisatellite missile to
serve first as a reconnaissance vehicle. Later, if satellite is
identified as unfriendly from telemetered information to headquarters, would destroy the satellite. A G.D.—Astronautics
proposal.

SINS (Navy/ONR)

Ship's Inertial Navigation Systems. Installed on first five
POLARIS submarines. NAA—Autonetics.

SIOUX

Research vehicle to be used for recovery experiments. R&D.

SJOROBOT (Sweden—Navy)

Torpedoboat model of ROBOT 315, surface-to-surface tactical
missile. See ROBOT 315.

SKEET (Air Force)

Antimissile missile.

SKIPPER (Navy)

Antisatellite weapon. Vertical probe or space mine which would
place cloud of small shot in a satellite's path. Satellite's high
speed would cause the shot to destroy it. Under study.

SKOKIE 1, 2 (Air Force)

Rocket-powered, air-launched parachute test vehicle. Program
to check rocket-powered sled data in order to develop recovery

systems for missiles and target aircraft. Length, 32 ft. Weight, 2500 lb (SKOKIE 1); 3000 lb (SKOKIE 2). Speed, approximately Mach 1 (SKOKIE 1); about Mach 2 (SKOKIE 2). Solid propellant. Cook Electric.

SKYBOLT (Air Force)

Air-launched ballistic missile (ALBM). Military designations, GAM 87A; XGAM-87A; WS-138A; WS-638A; WS-169A; WS-199B, C. Range, about 1000 miles. Hypersonic speed. Stellar-inertial guidance. Nuclear warhead. Planned for SAC air bases, RAF bases in England. May be launched from other aircraft than bombers, particularly transports and tankers. Program given cutback, but increased government emphasis has revived it. To be operational about 1964. Douglas, prime; Northrop-Nortronics, guidance; Aerojet, propulsion; G.E., re-entry vehicle.

SKYDART I, II (Air Force)

Rocket-powered target. Solid propellant. Expendable. Military designation, TDU-12/B. Curtiss-Wright.

SKYHOOK (Navy)

Project, part of ICEF, aimed at measuring high-energy cosmic rays. Three plastic balloons filled with 10 million ft^3 of helium and carrying 2900 lb of scientific equipment, were launched from the U.S.S. Valley Forge in January 1960. First balloon, SKYHOOK BRAVO, rose to 116,000 ft, but strong winds necessitated cutting off after 5 hours. Second launch, BRAVO RE-FLY, made 30 January 1960; balloon went to 113,500 ft, then lost altitude from ballast troubles. Third balloon launch, SKYHOOK CHARLIE, attempted 31 January 1960, but tore free and collapsed. Twelve more balloons scheduled to be launched from Fort Churchill to measure the charge distribution and energy spectrum of the heavy nuclear components in primary cosmic rays. NASA to pay for 11 shots and ONR for one.

SKYHOOK BRAVO

See SKYHOOK.

SKYHOOK CHARLIE

See SKYHOOK.

SKYHOOK RE-FLY

See SKYHOOK.

SKYLARK
A British sounding rocket fired successfully during the past few years in Woomara, Australia. Firings include four fired for NASA to measure stellar nebula fluxes of the southern continent. Six photometers used to measure flux between 2000—3000 A, and four other units with two ion chambers will measure conditions between 1000—2000 A. Bristol.

SKYLINER
Term suggested by Aerojet-General for vehicles used in future space travel.

SKYROCKET (Navy/NACA)
World's first aircraft to exceed Mach 2 in level flight (November 1953). Also in 1953 reached record altitude of 83,235 ft. Launched from bombers. Used same 6000-lb-thrust liquid-propellant rocket motor as the X-1. Military designation, D-558-II. Douglas.

SLAM (Air Force)
Supersonic Low-Altitude Missile. Military designation, WS-655A. Low-level nuclear missile. Supersonic speed, global range. Would use nuclear ramjet propulsion system similar to PLUTO. Program cost estimated at $500 million. Presently budgeted at about $7 million annually. Ling-Temco Vought, airframe; Lawrence Radiation Laboratory, reactor; Marquardt, propulsion.

SLC
Simulated Linguistic Computer. Georgetown University.

SLIM (Navy)
Submarine Launched Inertial Missile. Underwater-to-air-to-underwater missile. G.D.—Astronautics (Convair).

SLIM JOHN (Army)
Advanced model of HONEST JOHN. Douglas, prime.

SLOMAR (Air Force)
Space LOgistic Maintenance And Rescue. Studies of manned logistics spacecraft to serve operational space systems. Study contracts to industry—Norair, Martin, Douglas, G.D.—Astronautics, G.E., Lockheed-California (CALAC). Completed August 1961. SSD program.

SLRI
    Shipboard Long-Range Input. Supplement to ALRI (airborne
    long-range input) of the SAGE system which would add the
    capability for supplementary surface warning of airborne at-
    tacks. SLRI proposal request may be issued by the Air Force
    in late 1961.

SLV (Navy)
    Soft Landing Vehicle. 8-ft, rocket-powered vehicle with 1300-lb
    thrust. Capable of soft-landing men and instruments on the
    moon. A NOTS concept.

SLV-4
    See VANGUARD II.

SM
    Strategic Missile.

SM.20 (France)
    Surface-to-surface guided missile developed from the CT.20
    jet target drone. For use in coastal defense against naval tar-
    gets. Range, 155 miles. Warhead capability, 550 lb. See also
    MM.20 and R.20.

SM-65
    See ATLAS.

SM-68, SM-68B
    See TITAN.

SM-73
    See GOOSE.

SM-75
    See THOR.

SM-78
    See JUPITER.

SM-80
    See MINUTEMAN.

SMART (Air Force)
    Studies directed by Air Force Space Systems. Feasibility of
    manned satellite maintenance and repair techniques.

SMOKEY JOE (Air Force)
Identification of missile and space-vehicle launchings by infrared and other signatures through use of U-2 aircraft. Follow-on
to LOW CARD. Canceled.

SNAP I (AEC/Air Force)
Systems for Nuclear Auxiliary Power. Program to develop
nuclear electric power for space vehicles. Purpose: to provide
electrical power for the MIDAS system. Second prototype,
SNAP I-A scheduled to be tested in late 1961. For propulsion
power, see ROVER, KIWI. Martin.

SNAP I-A (AEC/Air Force)
125-w isotope power system for space use. Martin.

SNAP II (AEC/Air Force)
First demonstration reactor was called SNAP Experimental
Reactor (SER); final design called SNAP Developmental Reactor
I (SDR-I). SNAP II based on these models. 3-kw reactor unit
for space use. Purpose: to replace the equivalent of one-half
million-lb chemical batteries, and to be cheaper to operate.
Needed for CENTAUR and SATURN. NAA—Atomic Interna-
tional.

SNAP III (AEC)
Radioisotope-fueled thermoelectric generator weighing only
5 lb. Could be used to power a satellite transmitter for extended periods. Eight SNAP IIIs have been built, three fueled.
Martin.

SNAP IV (Navy)
Reactor. Classified. NAA—Atomics International.

SNAP VII AB (Coast Guard)
5- and 30-w isotope power system for use as floating and shore-
based navigation aid. Martin.

SNAP VII CD (Navy)
5- and 30-w isotope power system for Arctic weather stations.

SNAP VIII (NASA)
30- and 60-kw reactor unit for space propulsion and auxiliary
power supply. Should be ready by 1965. Would operate for at
least a year in flight. In the future, could be lofted in SATURN-
boosted spaceships. Weight, 15 tons or more. Still later, could

be placed in a space probe to Mars. Promising as a basic propulsion system as well as auxiliary. NAA-Atomics International/Aerojet-General.

SNAP IX (Navy)
10-w isotope power supply. Martin.

SNAP X (Air Force)
500-w reactor unit for classified space use. NAA-Atomics International.

SNARK (Air Force)
Jet-powered surface-launched, air-breathing ICBM. Length, 74 ft. Weight, 36,000 lb. Range, 5500 miles. Speed, Mach 0.9. Solid propellant. Thrust, 79,000 lb. Celestial inertial guidance. Assigned to SAC. Engine: X-226. Phasing out. Norair, prime; Northrop, guidance; Pratt & Whitney/ABL, propulsion.

SNOOPER
Low-thrust (1 to 2 oz) ion-propelled space probe which would use a big booster. Proposal by Rocketdyne.

SNOOPER (Army-Signal Corps)
Also called PEEPING TOM. All-weather surveillance drone. Propeller-driven, but launched by solid-propellant JATO units. Length, 15 ft. Weight, 1000 lb. Range, 100 miles. Interchangeable nose cones for various missions. Zero-length launch. Recoverable. Republic.

SNOWBALL (Air Force)
Series of DISCOVERER tests begun in 1959. Continued in 1961, making general systems checkout and setting up base-line indexes for any DISCOVERER biosatellite flight. See also BOSS.

SNOW GOOSE (Canada)
Code name given the two-year research in the Canadian BLACK BRANT rocket. Design by CARDE. Solid-fuel research and configuration evaluation by Canadair—Guided Weapons.

SNPO (AEC/NASA)
Space Nuclear Propulsion Office. Located in Germantown, Md. An organization established jointly by NASA and AEC to conduct the nuclear propulsion programs, including ROVER.

SOCOM (Air Force)
SOlar COMmunications system. Optical space communications systems using solar radiation for transmission. Expected to weigh about 50 lb. In R & D. Electro-Optical Systems, Pasadena, Calif.

SOFAR
SOund Fixing And Ranging. A bomb ejected from a missile or satellite payload as it strikes the water; used as a means of detecting and locating the payload. Used to permit radar fixes as for recovery of capsules, e.g., DISCOVERER and MERCURY capsules at sea.

SOLAR (Air Force)
Research vehicle conducting experiments at very high altitude. Eventually to use solar energy. In R&D.

SOLARIS (Navy)
Submerged Object Locating And Retrieving Identification System. TV-guided unmanned cable-operated vehicle capable of hauling loads up to 5000 lb in water to the surface. Designed originally to retrieve experimental Navy torpedoes and mines. Vitro Laboratories.

SOLO (Army)
Selective Optical Lock-On. Developed for missile guidance. May be applicable to guns aimed from Army tanks at moving targets. Chicago Aerial Industries has feasibility study.

SOMNIUM (ARPA)
In-house satellite project. Classified. Mentioned in Congressional testimony as a Pacific Missile Range project.

SOR
Specific Operational Requirement.

SOR-161
Air Force tactical range missile.

SOS
A compiling system for computers.

SPACE CANARY
Project to send instrumented monkey with human astronaut to warn if any dangerous environmental changes take place in the space capsule.

SPACECRAFT I
   See SPUTNIK IV.

SPACECRAFT II
   See SPUTNIK V.

SPACECRAFT III
   See SPUTNIK VI.

SPACECRAFT IV
   See SPUTNIK IX.

SPACECRAFT V
   See SPUTNIK X.

SPACECRAFT VI
   See SPUTNIK XI.

SPACECRAFT VII
   See SPUTNIK XII.

SPACE FERRY
   Three-stage vehicle proposed by Hughes and Lockheed, 1959, to take four men on a round trip to 300- to 500-mile altitude. Vehicle would have 1.3 million-lb thrust.

SPACE LAB
   Proposed by several companies, 1959. Northrop's conception: An ever-orbiting platform. Bullet-shaped vehicle with contemporary engines and attitude control jets. Crews ferried to platform by MERCURY-type capsule. Two men, 1-week tour. Purpose: to measure earth's atmosphere. Lockheed's conception: Wheel-shaped laboratory, 94-ft diameter. Capable of taking 10 men in 500-mile earth orbit. Nuclear.

SPACELINER
   Term suggested by Aerojet-General for vehicles used in future space travel.

SPACEMANPACK (Air Force)
   Small self-maneuvering back-pack propulsion system. Designed to allow mechanics to maneuver outside space vehicle while orbiting. Mock-up of wood and plastic to be built by 1962. Ling-Temco Vought.

SPACE METHUSELAH
    See VANGUARD I.

SPACE PLANE
    See AEROSPACE PLANE.

SPACER
    Classified project which will carry environment-measuring payloads. See also PIGGYBACK.

SPACETRACK (Air Force)
    National Space Surveillance Control Center. Military designations: SS-496L, WS-496L. Established to receive, analyze, and catalog orbital data from SPASUR. Now part of NORAD. See SPASUR.

SPACEWARN
    Special world communications net established during IGY. Provides scientists with orbital characteristics, payload weights, types of instrumentation, and experimental objectives.

SPAD (ARPA/Air Force)
    Satellite Protection for Area Defense. Part of DEFENDER. Two to three thousand satellites in random orbit at 200-nm altitude. Each satellite carries one to six missiles. Estimated to be a multimillion dollar AICBM system. Convair, Boeing, and Thompson Ramo Wooldridge to develop.

SPADATS
    SPAce Detection And Tracking System. A computer used to inform the NORAD commander of unidentified flying objects in space. In operation near Cole Springs at the NORAD Combat Operations Center. Budget for FY 1961, approximately $3 million; FY 1962, $30 million. Philco.

SPAEROBEE (Air Force)
    Upper-atmosphere research rocket. Military designation, AF-AA10.02. Altitude, 260 miles. Operational. Aerojet-General.

SPAR (Navy)
    Seagoing Platform for Acoustic Research. A stable platform in form of tube, used for mounting oceanographic instruments, e.g., those used in sonar and acoustic research. To be unmanned, but tended, and will operate in the Atlantic. Companion vessel FLIP to be manned and operated in Pacific.

SPAROAIR (Navy)
Two-stage rocket with two SPARROW missile motors. Unguided. Lofts 35-lb payload to 64 miles. Ten flight tests scheduled.

SPARROW 1 (Navy)
Air-to-air antifighter and antibomber. Solid propellant. Length, 12 ft. Weight, 300 to 350 lb. Beam-rider guidance. Speed, over 1500 mph. Range, 5 to 8 miles. Obsolete. Terminated December 1958. Sperry-Farragut.

SPARROW 2 (Navy)
SPARROW 1 with advanced guidance system. Discontinued. Douglas.

SPARROW 3 (Navy)
Air-to-air missile replacing the obsolete SPARROW 1. Military designation, AAM-N-6. Length, 12 ft. Launch weight, 400 lb. Radar-homing guidance. Solid-liquid propellant. HE warhead. Deployed with Navy aircraft and with Marine units. Operational, 1958. Raytheon, prime and guidance; Aerojet-General/Thiokol, propulsion.

SPASCORE (Navy)
System designed to show past, present, and future positions or paths of any satellite or number of satellites. Also capable of showing the predicted re-entry point of any satellite near the end of its life span and predicting the point of impact and final trajectory of a predetermined re-entry.

SPASUR (Navy)
SPAce SURveillance. System capable of detecting, tracking, and determining orbit of all satellites within its range that cross the detection line, whether the satellite is transmitting signals or is silent. Part of SHEPHERD. See DARK FENCE. NRL/ARPA.

SPAT
Silicon Precision Alloy Transistor. High beta and very low saturation voltage. Philco.

SPEAR
Solid rocket developed by Lockheed Propulsion (formerly Grand Central Rocket). Length, 14 ft. Diameter, 36 in. Weight of propellant, over 6500 lb. Thrust, 53,000 lb. Tested for capability of propellant to function under extremely poor environmental conditions.

SPEEDBALL
Target rocket for testing NIKE-ZEUS.

SPG-55 (Navy)
TERRIER guidance radar.

SPIN (ABMA)
Gyroscope expected to be five to ten times more accurate than existing gyros. Cryogenic techniques are used. G.E.

SPTV
Supersonic Parachute Test Vehicle. Rocket-boosted vehicle designed to provide a parachute test environment. Speed, to Mach 2. Dynamic pressures, to 4000 lb/ft^2. Length, over 22 ft. Weight, loaded, 2400 lb. Chute folded inside second stage of the vehicle. Developed by Sandia.

SPU (Navy)
Self Propelled Underwater plastic atomic missile. Navy frogman can straddle missile and steer toward target. After frogman has leaped off, robot mechanism takes over. Aerojet-General.

SPUD-1
TRW solar power unit.

SPUR (AEC/Air Force)
Space Power Unit Reactor. 300-kw reactor which begins to approach power levels necessary for large manned spacecraft. Has a 1-Mw potential. Study began in 1960. Possible that prototype tests could be made by 1966 if given funds. Design and development contract to Garrett AiResearch.

SPUTNIK I (U.S.S.R.)
Data-collecting experimental satellite launched 4 October 1957. Estimated length, about 60 ft. Launch weight, 184 lb. Booster thrust, approximately 600,000 lb. Estimated speed, 16,000 to 18,000 mph. Perigee, 142 miles; apogee, 560 miles. Initial orbit period, 96.2 min. In orbit 94 days, to 4 January 1958.

SPUTNIK II (U.S.S.R.)
Data-collecting experimental satellite launched 3 November 1957. Estimated length, approximately 80 ft. Launch weight, 1120 lb. Booster thrust, approximately 600,000 lb. Speed, 15,000 to 18,000 mph. Perigee, 140 miles; apogee, 1038 miles. Initial orbit period, 103.7 min. Discovered solar influence on the

density of the upper atmosphere through irregular acceler-
ations and delays. Carried instruments and dog Laika. In orbit
to 14 April 1958, or 163 days.

SPUTNIK III (U.S.S.R.)
Data-collecting experimental satellite. Estimated length, about
70 ft; 6 ft across base. Launch weight, 2950 lb. Speed, 14,640
to 18,825 mph. Perigee, 141 miles; apogee, 1168 miles. Initial
orbit period, 105.95 min. Carried instruments and solar batteries. Discovered information on extent and intensity of radiation in the exosphere; also on gas traces and ion concentration.
In orbit 691 days.

SPUTNIK IV (U.S.S.R.)
Also called SPACECRAFT I; KOSMICHESKIJ KORABL-SPUT-
NIK I. Experimental satellite launched 15 May 1960 by multistage rocket. Launch weight, 10,000 lb. Payload included
"dummy" astronaut, life support equipment, tape recorders,
retrorockets, and scientific instruments. Orientation fault in
orbit prevented re-entry. Perigee, 185.5 miles; apogee, 228.7
miles. Speed on burnout of last stage, 17,300 mph.

SPUTNIK V (U.S.S.R.)
Also called SPACECRAFT II; KOSMICHESKIJ KORABL-SPUT-
NIK II. Experimental satellite with objective of testing pressure
capsule and recovery equipment, launched 19 August 1960.
Capsule contained two dogs, and also instrumentation to investigate heavy nuclei in the primary cosmic radiation and
ultraviolet radiation and radiation inside the pressure capsule.
Perigee, 190 miles; apogee, 502 miles. Capsule re-entered
successfully during eighteenth orbit.

SPUTNIK VI (U.S.S.R.)
Also called SPACECRAFT III; KOSMICHESKIJ KORABL-SPUT-
NIK III. Experimental satellite launched 1 December 1960.
Orbited two dogs and burned up on re-entry during eighteenth
orbit.

SPUTNIK VII (U.S.S.R.)
Experimental satellite launched 4 February 1961. Payload
weight, 14,293 lb. Apogee, 204 miles; perigee, 139 miles. Improved multistage booster. Reported take-off thrust, 1 million lb.

SPUTNIK VIII (U.S.S.R.)
Venus probe launched 12 February 1961. Diameter, 80 in. Payload weight, 1419 lb. Aphelion, 94.6 million miles; perihelion,

66.7 million miles. Carried instrumentation to measure cosmic rays, magnetic fields, charged particles of interplanetary gas, and corpuscular sunbeams. Also recorded temperature control, micrometeor impacts, altitude control, and stabilization systems.

SPUTNIK IX (U.S.S.R.)
Also called SPACECRAFT IV; KOSMICHESKIJ KORABL-SPUTNIK IV. Experimental satellite launched and successfully recovered 9 March 1961. Carried dog Chernuska and other biological specimens. Payload, 10,360 lb. Apogee, 154 miles; perigee, 114 miles.

SPUTNIK X (U.S.S.R.)
Also called SPACECRAFT V; KOSMICHESKIJ KORABL-SPUTNIK V. Experimental satellite orbited 25 March 1961. Carried dog Zvesdochka and other biological specimens. Recovered successfully during first orbit. Payload, 10,350 lb. Apogee, 153 miles; perigee, 111 miles.

SPUTNIK XI
See VOSTOK I.

SPUTNIK XII
See VOSTOK II.

SQUID (ONR)
Program of basic research in missile and space propulsion sponsored by ONR. Incorporated in 1949. To be administered by Princeton University until 1 October 1962 when it will be transferred to the University of Virginia. Funded at about $550,000 yearly.

SR-126
See DYNA-SOAR.

SR-150
See ANP-NUCLEAR ROCKET.

SR-178
See CSAR.

SR-181 (Air Force)
Strategic Orbital System Study. Earth-orbital military satellite. Study completed and evaluated in 1959.

SR-182 (Air Force)
Study on requirements for orbiting vehicles beyond orbit of the moon. Entails use of missiles, satellites, maintenance and logistic vehicles, nuclear propulsion, and lunar bases. Study contracts to G.M., Westinghouse, and Douglas.

SR-183 (Air Force)
Study relating to establishment of a lunar observatory. Study completed and evaluated in 1959.

SR-192 (Air Force)
Strategic Lunar System Study. Lunar space system possibilities studied and evaluated in 1960.

SR-193 (Air Force)
Study relating to use of the moon as a strategic military base.

SR-199 (Air Force)
Third-generation ICBM; to be operational by 1970. To counteract antimissile missiles. MINUTEMAN follow-on. Douglas/NAA/Aerojet-General.

SR-17527
See MTSS.

SR-17530 (Air Force)
Design criteria for automatic test and checkout systems.

SR-17532 (Air Force)
Study requirement for permanent lunar base.

SR-17543 (Air Force)
Comprehensive study of bomber and air-to-surface missile combinations.

SR-19782 (Air Force)
Study requirement for small ICBM.

SR-19786
See ASP.

SR-49759 (Air Force)
Concepts of recovery of personnel and equipment from orbit or space.

SR-79811 (Air Force)
   TAC VTOL fighter.

SR-79821 (Air Force)
   Study requirement for earth-orbital weapon system (low orbit).

SR-79822 (Air Force)
   Study requirement for advanced earth-orbital weapon system
   (high orbit). Follow-on study to SR-79821. Investigates weapon
   carrier spacecraft at 10,000-mile altitude.

SR-79998
   See SATIN.

SR-89774 (Air Force)
   Recoverable Booster Study. Carrier vehicle recovery systems.
   Study completed and evaluated in 1959.

SRBM
   Short-Range Ballistic Missile.

SRLD
   Small Rocket Lift Device. Rocket belt under development by
   Bell Aerosystems with which wearer can attain controlled free
   flight over the ground. Belt consists of a twin jet, hydrogen
   peroxide propulsion system mounted on a fiberglass corset.
   A motorcycle-type hand throttle controls the rate of climb and
   descent.

SRN-2 (Britain)
   Air cushion vehicle designed to carry 54 to 66 passengers or
   100 standing soldiers over water. Weight, 27 tons. Size, 60 ×
   29.5 ft. Will cruise at 70 knots at 1 to 1.5 ft above surface of
   water. Test scheduled for 1962. Saunders-Roe.

SS-10 (Army)
   Surface-to-surface antitank missile. Range, 5200 ft. Speed,
   180 mph. Length, 34 in. Launch weight, 33 lb. HE warhead.
   Solid propellant. Replaced DART in U.S. Army. Now phasing
   out and being replaced by ENTAC. Nord Aviation, prime; G.E.,
   U.S. licensee.

SS-11 (Army)
   Antitank missile. Surface-to-surface or helicopter-to-surface.
   Length, 46 in. Range, 1650 to 11,500 ft. Speed, more than 400

mph. Wire guidance. Solid propellant. Armor-piercing HE warhead. Adopted by U.S. Army. To be manufactured in U.S. Nord Aviation, prime; G.E., U.S. licensee.

SS-12 (France—Army)
Surface-to-surface antitank missile. Length, about 6 ft. Weight, 150 lb. Range, more than 4 miles. Speed, Mach 1. Wire guidance. Solid booster. Nuclear or HE warhead. Advanced version of SS-10 and SS-11 sold to U.S. Army and to large number of NATO armies. Nord Aviation.

SS-413L
A distant early warning system for detecting hostile air-breathing vehicles approaching North American continent from the north.

SS-416L
See SAGE.

SS-425L
NORAD combat operations center.

SS-426L
See Q-4B.

SS-429L
See FIREBEE.

SS-433L (FAA)
Automatic weather observation and forecast system development program. FAA/United.

SS-460L
Weather reconnaissance support system.

SS-462L
Powered target support.

SS-465L
Strategic Air Command control system.

SS-466L
Electronic intelligence communication operations.

SS-470L
Communications satellite system.

SS-473L
Headquarters, United States Air Force operations control.

SS-474L
See BMEWS.

SS-496L
See SPACETRACK.

SS-497A
Miscellaneous NASA system work.

SS-498D
Strategic communications system.

SS-499C
Advanced system and space-vehicle studies.

SS-499D
Intelligence and reconnaissance system study.

SSBN (Navy)
Designation for a ballistic missile submarine.

SSBS-1 (France)
Surface-to-surface IRBM with 2300-nm range. Nuclear warhead. In R&D. Société d'Etude et de Réalisation d'Engins Balistiques.

SSD
See AFSSD.

SSM
Surface-to-surface missile.

SSM-5
See REDSTONE.

SSM-A-12
See LACROSSE.

SSM-A-14
See REDSTONE.

SSM-A-17
See CORPORAL.

SSM-A-23
    See DART.

SSM-A-25
    See NIKE-HERCULES.

SSM-A-26
    See SERGEANT.

SSM-N-8; SSM-N-8A
    See REGULUS I.

SSM-N-9A
    See REGULUS II.

SSN
    U.S. Navy designation for a nuclear attack submarine.

SSR-7790-19782
    See MIDGETMAN.

STAR
    Space Thermionic Auxiliary Reactor. Proposal for design of
    small nuclear conversion power units for space vehicles. Ther-
    mionic devices convert nuclear heat directly to electricity.
    Flight models could be built by 1966. Company-sponsored by
    G.E.—Atomic Products.

STARFIGHTER
    Designation, F-104. Fighter bomber aircraft developed by
    Lockheed-California. Has had great success in sales in several
    foreign countries, especially Germany (F-104G). Also called
    SUPER STARFIGHTER.

STARGAZER (Air Force)
    First manned balloon flight carrying telescope for astronomical
    exploration. Spring, summer 1961. AF Aeronautical Systems
    Division/M.I.T./Cambridge Research Laboratory.

STAR OF HOPE
    Star-shaped satellite proposed by Rep. Victor Anfuso for
    launching in December 1961. Purpose: to broadcast peace
    messages in various languages from world leaders. Cost,
    approximately $15 million.

START
Space Transport And Re-entry Tests. Re-entry test vehicle used proven solid rockets; to cost only one-fourth as much as liquid-fueled boosters. Basic rocket a Thiokol SERGEANT engine. Multistage vehicle capable of lofting test vehicles faster and higher than existing best systems. Boeing proposal.

STATALTEX (France)
Experimental high-altitude vehicle for ramjet propulsion tests.

STEER (Army/Air Force)
Strategic Polar Communications Satellite for SAC. A former project under ADVENT. Canceled.

STEP
Space Terminal Evaluation Program. Preliminary design studies made by Martin on an orbiting space station in cislunar space. Would use ICBM boosters until large space boosters are available. First space station would support three men. Later ones would be modified for 50 men. Estimated lifetime of station, 10 years. Company-funded.

STEPS (Air Force)
Solar Thermionic Electric Power System. Lightweight source of kilowatt power for advanced satellites. System composed of a parabolic reflector which focuses the sun's rays on a generator made up of many thermionic converters and several subsystems. First system to deliver 500 w continuously; later systems, 3 to 10 kw.

STOL
Short Take-Off and Landing.

STORC
Self-ferrying Trans-Ocean Rotary-wing Crane. Proposed giant flying crane helicopter capable of conversion to an aircraft by removing a rotor and positioning both rotors in such a way that they would serve as a wing. Wing-tip engines are then used in flight. Hiller.

STORIC (Air Force)
Surface-to-surface missile. Vehicle composed of a THOR IRBM plus upper staging as an ICBM, calling for use of storable propellants. A Douglas proposal in 1959.

STRAC
STRategic Army Command. Now composed of three airborne or air-transportable divisions (115,000 men). Has responsibility for cases of general and limited war. Depends on Air Lift Command for long-range movement. DOD has combined this with TAC and dubbed it the U.S. Strike Command.

STRAT-LAB (ONR)
Manned balloon flights making telescopic studies at high altitudes. ONR/JHU.

STRATOLAB HIGH (Navy)
High-altitude, manned balloon flight made 4 May 1961 by two Navy officers in modified MERCURY full-pressure suits. Carried bio-instrumentation. ONR/Winzen Research.

STRATOSCOPE I
A 12-in. balloon-borne telescope for astronautical observations. Developed by Martin Schwarzschild of Princeton University. Successful experiment. Six flights, 1957-59. Princeton University/Perkin-Elmer.

STRATOSCOPE II (ONR/NSF/NASA)
A 36-in. balloon-borne telescope for astronomical observations. Under development by Martin Schwarzschild at Princeton University. Will carry telescope to 80,000-ft altitude. Accuracy, 0.1 sec of arc over extended periods. Will permit observation of Venus, Pluto, Jupiter, and Saturn. Princeton University/Perkin-Elmer.

STREAMLINE (Air Force)
Code name used to designate proposed SATURN-boosted DYNA-SOAR. This might place man in orbit much faster than planned earlier, and could reduce total program costs by $300 million. Boeing.

STRIKE
Unified SAC-STRAC command.

STRIVE (ARPA)
Basic research concerning systems reliability. In study phase.

STRONG ARM (NASA/Army)
Five-stage research vehicle capable of lofting 15-lb payload to 1050-mile altitude. Length, over 56 ft. Weight, 7125 lb.

Liquid-solid propellant. Used to gather data on ionization decrease at altitudes of 1000 miles. University of Michigan.

SUBROC (Navy)
Underwater-to-underwater or surface-to-underwater missile with aerodynamic and hydrodynamic surfaces. Range, about 25 nm. Solid propellant. Nuclear or HE warhead. Scheduled to be installed aboard nuclear-powered submarines. Operational in 1961 or early 1962. Superseded RAT. Goodyear Aircraft, prime; Thiokol, propulsion; Kearfott, guidance.

SUM
Surface-to-underwater missile.

SUNFLARE (NASA)
Study of solar flares and solar temperatures made in a series of high-altitude shots. Boosters, NIKE and AEROBEE.

SUNFLOWER (NASA)
Solar auxiliary power system, designed to generate 3000 w of electrical power for 1 year continuously. Has potential use in satellites and lunar and planetary probes. Two 3000-w systems to be demonstrated in 1964. Prototype systems by Tapco/Thompson Ramo Wooldrige.

SUNRAY
See GREB.

SUNRISE (ARPA)
Studies of advanced weapon concepts, not currently in research, to be undertaken in SUNRISE program. Associated with ARCADE, which is concerned with specific research.

SUPER BOMARC
Advanced model of BOMARC missile. Radar-command—active-homing guidance. Launch weight, about 15,000 lb. Range, 475 miles. Altitude, 100,000 ft. Speed, Mach 4.

SUPERSONIC III
Solid-propellant research vehicle. Length, 35 ft. Weight, 15,000 lb. Capable of lofting 2000-lb payload to 70,000-ft altitude. In R&D.

SUPER STARFIGHTER
See STARFIGHTER.

SUPER TALOS (Navy)
Advanced version of TALOS missile. Has antimissile capability. Bendix.

SUPER TARTAR
See TYPHON (medium range).

SURFACE EFFECT SHIP
Also termed a ground effect machine (GEM). For use in skimming the surface of water or ground.

SURVEYOR (NASA/JPL)
Lunar soft-landing vehicle. Spacecraft to make three-point, 10-mph, soft landing on moon. Payload weight, 750 lb, of which 250 lb are scientific instruments, including spectrometer, lunar surface drill, Sterox color TV cameras. Modified B version of SURVEYOR to make orbital reconnaissance mission to moon in late 1964. Considered that more than the original seven launchings will be made. Estimated cost of program, $175 million, plus R&D. Cost per unit, approximately $15 million. Cost per launch, approximately $25 million.

SUZANO (ARPA)
Study program for a space platform to provide an advance base for space missions. Canceled in October 1959; reactivated, and now in study phase.

SWALLOW (Army)
Solid-propellant, JATO-boosted surveillance drone. Turbojet-sustained. Weight, 3500 lb. Range, about 200 miles. Canceled by Army. May be bought by Navy to be used as targets. Fairchild.

SWEEP (NBS)
Structures With Error Expurgation Program. A data-processing system for automatic handling of data for the U.S. Patent Office.

SX-A5 DEVELOPMENT (Great Britain—Navy)
Small guided weapon intended to replace light antiaircraft guns on surface ships. Possibly similar to MALKARA. Outgrowth of SX-A5. Short Bros./Harland.

SYNCOM (NASA)
System of three small active satellites moving in a figure-eight pattern, 33° N and S of the equator in 22,300-mile orbit. Payload weight, 50 lb. First launching scheduled for late 1962

by DELTA rocket. Small solid rocket attached to the satellite to give the added push to put it into orbit since the DELTA is not powerful enough. Satellite to be available to commercial and government communications organizations around the world. Will relay telephone and telegraph, but not TV. $4 million contract, September 1961, to Hughes for three flight units. Launch costs estimated at approximately $3 to $3.5 million each.

SYSTOS (Air Force)

System offices ordered by Air Secretary Eugene M. Zuckert to handle high-priority projects. Purpose: to shorten the time for decisions on major weapon systems through taking up the problems encountered by AFSC immediately with the appropriate Pentagon board rather than waiting for ruling by the Weapons Board, the Air Council, USAF Chief of Staff, and the Air Secretary. Ten offices to be created initially; more may be created later. SYSTOS to handle MINUTEMAN, ATLAS, TITAN, and the B-70.

T

T-1 (U.S.S.R.—Army)

Russian name, Pobeda (Victory). Military designation, M-101. Surface-to-surface mobile IRBM. Length, 62 ft. Weight, nearly 40,000 lb. Range, 600-775 miles. 2550-lb nuclear warhead. Radio-guided. 77,000-lb thrust by LOX/hydrocarbon fueled booster. Copy of German A-4 missile. Reportedly over 3000 in use, with a few given to Red Chinese troops. First seen in Moscow, 1956.

T-2 (U.S.S.R.—Soviet Missile Command)

Surface-to-surface missile, supposed to be the first Soviet rocket used to test H-bomb warhead. Military designation, M-103. Reported length, 65.5 ft, 80 ft, and 100 plus feet. Liquid-fueled booster. Range, 1800 miles. Speed, about Mach 13. Based in Soviet satellite countries. Reportedly, 700 in service. T-2 developed from plans of German rocket A-4/A-9/A-10 combination found by the Russians in Nordhausen after World War II.

T-3 (U.S.S.R.—Air Force)

Surface-to-surface missile, 500,000-lb thrust booster. Radio-inertial guidance. Range, 5000 miles. Speed, 15,000 mph. Apogee, 280 miles. Nuclear or HE warhead. Missile reported to be able to hit within 10 miles of target. Also reported that 50 were produced in 1959 and 1200 to be built by end of 1963. Operational.

T-3A (U.S.S.R.—Air Force)
Surface-to-surface liquid-fueled ICBM. Model A: Range, 6000 miles. Thrust, 525,000 lb. Model B: Range, 7500 miles. Speed, 15,000 mph. Thrust, 700,000 lb. Model A production believed stopped in favor of the advanced version. Model B believed to be a prototype for one used in the T-4A antipodal missile. Nuclear warhead. Operational.

T-3 Mk 1 (U.S.S.R.—Soviet Missile Command)
Surface-to-surface ICBM, in operational service in 1959. Military designation, M-104. Length, more than 88 ft. Weight, 176,000 lb. Range, 4950 miles. Speed, Mach 20. Liquid-fueled. 2200-lb thermonuclear warhead. Radio-inertial guidance. Guidance in second stage.

T-3A Mk 1 (U.S.S.R.—Soviet Missile Command)
Surface-to-surface ICBM in operational status. Length, 91.5 ft. Weight, 175,000 lb. Range, 6200 miles. Speed, Mach 21+. Liquid propellant. 1100-lb thermonuclear warhead. Programed guidance. An advanced T-3 fired from fixed bases. In production, and being considered for hardened sites.

T-3 Mk 2 (U.S.S.R.—Soviet Missile Command)
Surface-to-surface ICBM under development. Length, 108 ft. Weight, 352,000 lb. Range, 6500 miles. Speed, Mach 20+. Liquid propellant. 2500-lb thermonuclear warhead. Inertial guidance in second stage. Used to launch satellites into space. First used in 1957.

T-3A Mk 2 (U.S.S.R.—Soviet Missile Command)
Surface-to-surface ICBM. Liquid propellant. Length, 101.5 ft. Weight, 395,000 lb. Range, 6800 miles. Speed, Mach 22.1. Has 1250-lb thermonuclear warhead. Programed guidance. Fired from fixed bases. Operational. In mass production. Reported to be the booster used for the Soviet man-in-space shot.

T-4 (U.S.S.R.—Army)
Surface-to-surface IRBM. Military designation, M-102. Two-stage, both liquid-fueled. Length, possibly 50 to 55 ft. Weight, 35 tons. Payload, 1800 lb, either nuclear or HE. Range, 1000 miles. Speed, 9500 mph. Some of its configurations believed to be in the T-4A upper stages. Developed from the German A-10. Stubby wings on some versions. Experimental status. Not too successful, by report.

T-4A (U.S.S.R.—Air Force)
   Surface-to-surface antipodal missile. First-stage powered by
   LOX/kerosene with 360,000-lb thrust. Range, 12,500 miles.
   Speed, Mach 19+. Altitude, 186.3 miles. Payload, 2400 to
   3100 lb. Radio programed. Similar to U.S. DYNA-SOAR boost-
   glide vehicle, stemming from the World War II German Saenger-
   Bredt design of the antipodal rocket bomber. Used in the Soviet
   man-in-space programs. In RDT&E.

T-5 (U.S.S.R.—Army)
   Surface-to-surface solid-fueled missile designed for firing
   from multiple launchers for heavy saturation of target. Range,
   50 to 100 miles. Speed, Mach 4+. Copy of German RHEINBOTE.
   Operational with Soviet Army in Eastern Europe.

T-5A (U.S.S.R.—Army)
   Surface-to-surface IRBM, believed to be solid-fueled and guided.
   Operational.

T-5B (U.S.S.R.—Army)
   Surface-to-surface missile. Length, 30 ft. Weight, 5750 lb.
   Range, 15 to 25 miles. Speed, Mach 2+. Unguided. Launched
   from tracked vehicle. Solid propellant. HE, possibly nuclear,
   warhead. Similar to HONEST JOHN. First seen in 1957.
   Standard in all Soviet armored units. Operational but not in
   production.

T-5C (U.S.S.R.—Army)
   Surface-to-surface solid-propellant missile. Length, 25 ft.
   Weight, 4400 lb. Range, 25 miles. Speed, Mach 3+. 1100-lb,
   HE or nuclear warhead. No guidance. Introduced November
   1957. Fired from PT-76 amphibian tank chassis. Operational.
   New version now in production has warhead with different shape.

T-6 (U.S.S.R.—Army/Air Force)
   Surface-to-air solid-fueled missile. Has cluster of four solid
   boosters. Length, 22 ft. Weight, 4000 lb. Range, 20 to 25 miles.
   Speed, Mach 2.5. Ceiling, approximately 60,000 ft. Operational.
   88-lb HE warhead. Warhead coasts to target on ballistic trajectory after finned nose cone separates by explosive bolts.
   Advanced version (T-6A) is radar-guided, believed to be operational.

T-7 (U.S.S.R.—Air Force)
   Surface-to-air liquid-fueled missile. Length, 30.1 ft. Weight,
   5050 lb. Range, 60 miles. Speed, nearly Mach 6. Ceiling, 100,000

ft. 407-lb HE warhead. Radar guidance. Originally a high-altitude research vehicle. Assembly plant near Moscow.

T-7A (U.S.S.R.—Army)
Surface-to-surface solid-fueled missile with controllable rear fins. Length, almost 28 ft. Weight, 8800 lb. Range, 50 to 90 miles. Speed, Mach 4+. Radio-command guidance. 1770-lb, HE or nuclear, warhead. Mounted on KW-85 tank chassis and launched from vertical position. Comparable to U.S. CORPORAL.

T-8 (U.S.S.R.—PVO-Air Force)
Surface-to-air, air-to-air solid—liquid-fueled missile. Range, 15 to 25 miles. Speed, Mach 2.5. HE warhead with proximity fuze. Operational since 1954. 98% kill rate claimed against J-1 target drones. Deployed in field batteries, six launchers to a battery.

T-9
See KOMET 1.

T-10
See KOMET 2.

T-14
"Jeep of the deep." An underwater vehicle designed for oceanographic research. Made of lightweight aluminum. Weight, 180 lb, fully equipped. Can dive, loop, climb, and roll under water. Length, 9.5 ft. Less than a foot wide. Operates at a constant speed of 3 knots for up to 2 hours at depths to 230 ft. Designed by Dr. Dimitri Rebikoff of Loral Electronics.

T-33 (U.S.S.R.—Army)
Surface-to-surface ICBM. New development. No details available.

T-33 (NASA)
Solid-propellant research vehicle. Operational.

T-210
See LOKI.

T-238
See BOLT.

T-273 (Army)
Weapon system developed by AOMC, Chemical Corps, and Ordnance Weapons Command.

TAC
Air Force Tactical Air Command. Hq., Langley AFB, Va. DOD
has combined this with STRAC under the U.S. Strike Command.

TACKLE (Army)
Advanced polar communications satellite. A former project
under ADVENT. Canceled. See ADVENT.

TAIFUN
German unguided barrage rocket. See LOKI.

TALANT
Thiokol Nuclear Development Center/Allison Division/Linde
Division, Union Carbide/and Nuclear Development Corp. Team.
A group organized for handling major contracts, e.g., KIWI
reactor.

TALKING BIRD (Air Force)
Communications system integrated into a single package which
can be rolled into transport plane and set up quickly in remote
areas where communications are required. Two TALKING
BIRDs to be assigned to TAC, two to the Air Force-Europe,
and two to Pacific Forces.

TALOS (Navy)
Surface-to-air antiaircraft missile. Military designations:
SAM-N-6; XSAM-N-6; IM-70; RTV-N-6. Length, 30 ft (with
booster). Weight, 3000 lb. Range, 70 miles. Speed, Mach 2.5.
Beam-radar guidance; semiactive radar homing. HE warhead.
Operational. Engine: X-251. SUPER TALOS under development.
Outgrowth of BUMBLEBEE program. TALOS system improve-
ments have led to the TYPHON antimissile system. Bendix,
prime; Bendix/Sperry, guidance; Naval Propellant Plant, pro-
pulsion.

TALOS W (Army)
Land version of TALOS developed for Army. Competitor to
NIKE-HERCULES system. Initiated in 1953; canceled in July
1957. Bendix.

TARP
Tactical Airborne Recording Package. Purpose: to monitor the
operational performance of the armament control systems of
the FALCON. Consists of an oscillograph recorder and eight
plug-in extensions.

TART (Air Force/Navy)
Small countermeasures rocket. Solid propellant. In production 1960. Thiokol.

TARTAR (Navy)
Surface-to-air (sea-going) antiaircraft missile. Weight, about 1500 lb. Length, 15 ft. Range, 10 miles. Supersonic speed. HE warhead. Solid propellant. Operational aboard one cruiser, 1961. G.D.—Astronautics, prime; Raytheon, guidance; Aerojet-General, propulsion.

TARZON (Air Force)
Unpowered guided bomb, similar to AZON and RAZON, but much larger and more advanced. Weight, 6 tons. Carried by B-29, B-50, and B-36. TARZON project was begun during World War II, dropped in 1946, and revived for Korean War in 1948. Discontinued September 1951. Bell Aircraft.

TATM-1 (Japan—Army)
Surface-to-surface antitank missile. Length, 4.5 ft. Launch weight, 300 lb. TATM-2 also under development. Length, over 3 ft. Kawasaki Kokuki Kogyu Kabushi-iki Kaisha, prime.

TATTLETALE (Air Force)
Emergency communications systems. Tests made using SPAE-ROBEEs to determine possibility of using rockets to send communications over several thousand miles. Boosts 47-lb transmitter to 300-mile altitude. Project began 31 August 1960. Tests by Hughes.

TCBM (Air Force)
TransContinental Ballistic Missile. Designed to reach almost any target on earth from U.S. Probable booster: ATLAS or TITAN first stage with winged or aerodynamic payload. Range, 11,000 miles. In R&D. Boeing.

TD-165
Liquid rocket motor used with CORVUS. Thiokol.

TD-174
See LR44-RM-2.

TDDL (Air Force)
Time Division Data Link. A single multiplex communications system used by the AFSC to issue multiple missile/aircraft defense commands. To be used with BOMARC B.

TDMS (Air Force)
Telegraph Distortion Measuring System. Development, 1961. Radiation, Inc.

TDU (Air Force)
Series of tow targets.

TDU-6B (Air Force)
Airborne missile tow-target. Electronic Specialty.

TDU-12/B
See SKYDART.

TE-29
See RECRUIT.

TE-64
See XM29.

TE-77C
See XM38.

TE-316 (NASA)
MERCURY capsule retrorockets. Thiokol.

TEAK (Air Force)
Code name for nuclear bomb detonated over Johnston Island in the Pacific in August 1958. (Used with ORANGE.) Shots made marked geomagnetic disturbances and radio blackouts over thousands of square miles.

TEAL (Navy)
Air-launched rocket-powered drone. Military designation, XKDT-1. Altitude, 10 miles. Transonic speed. Emits flares to assist in visual tracking and to signal "near misses." Temco.

TEAS (Air Force)
Threat Evaluation and Action Selection. Project to collect information for development of a command and control center that might be used by Joint Chiefs of Staff after 1970. Improvement on concept of GLOBEQUICK. $1 million spent to FY 1961. Cambridge Research Laboratory. G.E. contributing.

TELCAR (NASA)
Shipborne system for tracking MERCURY capsule in the atmosphere. Canoga Electronics.

TELESAT
Two stationary telecommunications satellites in equatorial orbit. Altitude, 22,300 miles. Payload, 1000 lb. Should be capable of 1000 or more two-way voice channels by 1967. Booster, CENTAUR AGENA C. Lockheed proposal.

TELSTAR (U.S./Britain)
ATT experimental commercial satellite. Project to study capability of satellites to carry about 1000 telephone channels and one or two television channels; also to provide continuous 24-hr service for a system of telephone and telegraph stations and world-wide coverage in conjunction with present radio and cable links. Bell Labs building two satellites. Diameter, 34 in. Weight, 150 lb each. To be launched by THOR-DELTA. Cost, $6 million per launch, successful or not. Total program costs estimated at approximately $200 million. First shot scheduled for June 1962; second shot, October 1962. Backups available.

TENOC (National Academy of Science)
Ten-year program of oceanographic research recommended by National Academy of Science. Cost, over $1 billion.

TEPEE (Navy/ONR)
Also called THALER'S PROJECT. Uses phenomenon of ionospheric backscatter to obtain information about ionization caused by nuclear explosions or missile exhaust columns. Could develop into Navy early warning system.

TERNE III (Navy)
Surface-to-underwater ASW missile. Length, 6.5 ft. Weight, 265 lb. Range, more than 3 miles. Solid propellant. 105-lb Mk 7 depth charge (HE). Six missiles can be launched in 5 sec. Navy buying from Norway to equip two destroyer escorts. Kongsberg Vapenfabrikk, prime; Arma, systems integration.

TERRAPIN (Navy)
Economical, upper-atmosphere weather reconnaissance rocket. Capable of lofting a 6.5-lb payload to over 400,000 ft at 3800-mph peak velocity. Launchable from a SKYHOOK balloon, like FARSIDE or ROCKOON. Length, 14.6 ft. Weight, 224 lb. Discontinued. University of Maryland/Republic Aviation.

TERRIER (Army)
Land version of the surface-to-air Navy TERRIER. Program initiated May 1951; terminated September 1956.

TERRIER (Navy)
Surface-to-air antiaircraft missile for shipboard use. Military designation, SAM-N-7. Based on experimental LARK. TERRIER 1: Range, 10 miles. Speed, Mach 2.5. TERRIER 2: Range, 20 miles. Length, 24 ft (with booster). Beam-rider guidance. Operational since 1956. Advanced TERRIER, with 100% greater capability, deployed 1961. The frigate Dewey, first ship armed with the advanced TERRIER. See ASTER. Convair, prime; Reeves/FTL, Sperry, guidance; ABL, propulsion.

TFX (Air Force/Navy)
Multipurpose, single-seat tactical fighter based on a concept of variable-geometry lifting surfaces and an advanced turbofan powerplant. Military designation, F-111. Program to be managed by Air Force Aeronautical Systems Division, Wright-Patterson AFB, with full Navy participation. Ninety-day study contracts awarded in January 1962 to Boeing and Convair.

TGAM-83 (Air Force)
Missile used in training jet pilots in use of GAM-83, the Air Force version of Navy BULLPUP. Martin has $1.5 million contract to build 424 TGAM-83 missiles.

THALER'S PROJECT
See TEPEE.

THERMOSORB
Coolant blanket developed for aerospace vehicle applications. More than 90% water but with handling characteristics of a solid. Ling-Temco Vought-Astronautics.

THIAMAT (NACA)
Surface-to-air missile developed in 1946 by NACA. Designation, MX-570. Length, 14 + ft. Altitude, 13,000 ft.

THOR (Air Force/NASA)
Liquid-fueled, land-based IRBM. Length, 65 ft. Weight, less than 100,000 lb. Speed, over 10,000 mph. Range, more than 1500 miles. Thrust, 150,000 lb. Inertial guidance; ballistic. Engine: XLR 70 NA-9 MB-3. Operational. Deployed in British Isles. Used also as first stage in space boosters, e.g., THOR-ABLE, THOR-EPSILON, THOR-DELTA, THOR-AGENA. Douglas, prime; AC Spark Plug, guidance; Rocketdyne, propulsion; G.E., re-entry vehicle.

THOR-ABLE (Air Force/NASA)
Satellite launcher for NASA; also deep-space probe. Three stages. Consists of THOR plus modified VANGUARD upper stages. Launched 270 lb. Operational. Phased out. Douglas/STL.

THOR-ABLESTAR
See THOR-EPSILON.

THORAD (Air Force)
Advanced THOR booster, with four assist rockets attached to the airframe. May be used in ballistic and orbital shots of the RITA program. Also a surface-to-surface missile. Study, and R&D. Douglas.

THOR-AGENA A, B (Air Force/NASA)
Two-stage liquid-propellant booster capable of launching (A model) 300 lb into 300-mile orbit; (B model) 1600 lb into 300-mile orbit. Length: (A) 78.6 ft; (B) 86 ft. Gross weight: (A) 118,500 lb; (B) 123,000 lb. All-inertial guidance. THOR-AGENA A used in DISCOVERER program. THOR-AGENA B used or scheduled for use in the following programs: ALOUETTE (S-27); ANNA; DISCOVERER; ECHO II; OSO; POGO; and NIMBUS. Five, possibly ten AGENA B's to be bought by NASA by 1963, in addition to present five. Engine: XLR 81-BA-3; -5; -7; -9. Lockheed/Douglas.

THOR-DELTA (NASA/Air Force)
Liquid-solid propellant booster, using THOR as first stage. Capable of lofting 132 lb to 1000 miles or 500 lb into 300-mile orbit. Length, 92 ft. Weight, 112,000 lb. First shot unsuccessful attempt to loft ECHO I in 1960. Has orbited TIROS II and EX-PLORER X satellites; scheduled to orbit RELAY; S-3; S-6; S-16. Will replace the SCOUT in a NASA-British scheduled space firing. Also scheduled to launch SYNCOM satellites in late 1962 and ATT TELSTAR communications satellites in June and October 1962. Douglas, prime; Aerojet/ABL, upper-stage propulsion.

THOR-EPSILON (Air Force/Army/Navy/NASA)
Formerly THOR-ABLESTAR. Booster for earth-orbiting satellites. Consists of THOR plus modified THOR-ABLE second stage. Second stage has restart engine. Length, 79 ft. Gross weight, 119,000 lb. Used to launch TRANSIT and COURIER satellites. STL, prime; Rocketdyne/Aerojet-General/ABL, propulsion.

THOR-HUSTLER (NASA)
Former name for THOR-AGENA, satellite launcher used in DISCOVERER program.

THORIC (Air Force)
Intercontinental version of THOR. Liquid propellant. Study by Douglas.

THUMPER
Surface-to-air rocket. Range, 35 miles. Command guidance. Canceled. G.E.

THUNDERBIRD (Great Britain—Army)
Surface-to-air missile deployed with British Army combat troops. Solid propellant. Range, 25 miles. Speed, Mach 2. Radar-homing guidance. HE warhead. Advanced THUNDERBIRD with low-level capability and increased range entering production. Operational in 1959. English Electric Aviation.

THUNDERCHIEF (Air Force)
Designation F-105A. Fighter-bomber developed by Republic.

THUNDERSTICK
Antitank and antiaircraft solid rocket. Ground-launched. Length, 6 ft. Diameter, 3 in. American Rocket.

TIDDLE
Nickname for TDDL.

TIGERCAT (Great Britain)
Surface-to-air antiaircraft missile, designed to be launched from armored carriers and to replace conventional artillery. Outgrowth of SEACAT. In R&D. Under consideration by British Army. Short Bros./Harland.

TING-A-LING
Version of the MB-1 GENIE.

TINY TIM (Navy)
Air-to-surface rocket developed immediately after World War II. Pioneer in the air-to-surface field. NOTS.

TIROS (NASA)
Television and InfraRed Observation Satellite. Series of seven weather satellites. Average weight, 285 lb. Forerunner of NIM-

BUS and AEROS programs. THOR-DELTA and THOR-AGENA
B boosters. RCA/Army Signal Corps, prime.

TIROS I (NASA)
Also called PILLBOX. Designation, S-60-5A. Lofted from Cape
Canaveral 1 April 1960 by THOR-ABLE. Weight, 273 lb. Height,
19 in. Diameter, 42 in. Equatorial, circular orbit. Perigee,
428.7 miles; apogee, 465.9 miles. Inclination to the equator,
48.3°.

TIROS II (NASA)
Weather satellite launched from Cape Canaveral 23 November
1960 by three-stage THOR-DELTA. Dimensions similar to
TIROS I. Equatorial, circular orbit. Inclination to equator,
48.5°. Perigee, 387 miles; apogee, 453 miles. Equipped with
two television cameras, seven IR sensors, and instruments to
measure solar radiation. In orbit more than a year, relaying
billions of bits of IR data for measurement of earth's heat
balance. Also has transmitted more than 35,000 photographs
of earth's cloud cover.

TIROS III (NASA)
Also called HURRICANE HUNTER. Weather satellite launched
from Cape Canaveral 12 July 1961 on a schedule placing it in
daylight over the northern hemisphere during the September
hurricane season. Weight of satellite, 285 lb. Carries two wide-
angle TV cameras, taking "stills" and storing their electronic
images on tape during each orbit of the earth. Inclination to the
equator, 48°. Perigee, 460 miles; apogee, 505 miles. Purpose:
to spot hurricanes at their birth as new storm warning system.
However, such cloud patterns are still unknown to scientists.
They will, therefore, have to trace storms backward, after
they have occurred, to their beginnings through TIROS pictures.
TIROS and NIMBUS budgeted for $50 million, FY 1962. RCA/
Army Signal Corps, prime.

TIROS IV (NASA)
Weather satellite scheduled for launch by THOR-DELTA booster
during second quarter of 1962. Experiments in TIROS IV (and
remaining three of the series) expected to be similar to those
in the current TIROS III. Considered likely that the narrow-
angle camera will be omitted, however, and that two of the
final satellites will contain infrared subsystems. TIROS V and
VI shots scheduled for firing before NIMBUS and AEROS programs begin.

TISSUGLAS
All-glass insulating paper. Purpose: to solve unusual electrical, thermal, and electronic problems. Potentially, largest application is as base for electroluminescent lamps. Also manufacture of bonded strain gages and as a component for printed circuit materials. American Machine & Foundry.

TITAN (Air Force/NASA)
Surface-to-surface ICBM. Military designations: WS-107A; WS-107A2; SM-68. TITAN I: Length, approximately 90 ft. Weight, estimated at 200,000 lb. Speed, Mach 15. Range, 5500 miles. Thrust, first stage, 300,000 lb; second stage, 60,000 lb. Liquid propellant. Operational in 1961. Nine TITANs to a squadron. First squadron at Lawry AFB near Denver operational at end of 1961. TITAN II: Length, 102 ft; first stage, 70 ft. Range, 6300 miles. First stage thrust, 430,000 lb; second-stage thrust, 100,000 lb. Lift-off weight at launch, 150 tons. Operational in 1963. For use in lofting DYNA-SOAR. A three-stage version proposed by Martin would be a basic TITAN II with an Aerojet-General third stage, HYLAS-STAR. Payload capability, about 6 tons. Total thrust estimated at 750,000 plus lb. Martin, prime; Bell/Sperry, TITAN I guidance; AC Spark Plug, TITAN II guidance; Aerojet-General, propulsion; Avco, TITAN I re-entry vehicle; G.E., TITAN II re-entry vehicle.

TITAN III (Air Force)
A proposed booster using TITAN II with two solid 120-in.-diameter strap-on NOVA boosters. This would add 2 million-lb thrust and would provide additional thrust to DYNA-SOAR 2 and 3, pending development of F-1 engine. Would enable DYNA-SOAR to go into orbit or would provide greater warhead-carrying capacity than ICBM's. To be developed in FY 1962 and 1963. Martin.

TLRM-1 (Japan—Army)
Surface-to-air long-range missile. Speed, about 255 mph. Length, over 10 ft. Launch weight, about 560 lb. Possibly radar guidance. In R & D. Mitsubishi Shipbuilding & Engineering.

TM
Tactical Missile.

TM-61A; TM-61C
See MATADOR.

TM-76A; TM-76B
   See MACE.

TMA-1 (Japan—Air Force)
   Air-to-air supersonic missile. Length, about 10 ft. Launch
   weight, more than 200 lb. Solid propellant. MM-1 is test vehicle
   for the TMA-1. Mitsubishi Electrical.

TMX (Air Force)
   Experimental tactical missile. Range, 1000 to 1500 miles.

TOMAHAWK (Army)
   Light antitank missile, possibly successor to SS-10 and ENTAC.
   Study by Martin.

TONTO (NASA)
   Lunar package (weight, 300 lb) to be carried by 800-lb RANGER
   probe for hard landing on moon, 1961-62 time period. Carries
   instruments to record moon tremors and meteor impacts,
   measure radiation, and determine surface temperatures.

TOPAZE (France)
   Research rocket developed from the AGATE rocket, but smaller
   and more advanced. May be used as second stage for booster
   to launch 100- to 175-lb satellites in the space research program. First stage may be either EMERAUDE or SAPHIR. Third
   stage may be DIAMANT.

TOPSIDE SOUNDER (Canada/NASA)
   Former name for U.K.I, the S-27 satellite to be launched as
   part of NASA's S-51 under direction of Defense Research Telecommunications Establishment of Canada. Purpose: study of the
   ionosphere up to 700 miles. DeHavilland Aircraft of Canada.

TORES (ARPA)
   Research into the toxicological effect of new chemicals and
   metals on humans in order to ensure safe and economical handling. These chemicals and metals used primarily in new propellants, explosives, solvents, fuels, and lubricants. In study
   phase.

TORPEDO MARK 32 (Navy)
   Surface-to-underwater lightweight, acoustic-homing antisub-
   marine missile to replace torpedoes. Fired by surface ships.
   Philco.

TORPEDO MARK 35 (Navy)
An acoustic-homing torpedo for antisubmarine use. Fired by surface ships.

TORPEDO MARK 37 (Navy)
Antisubmarine acoustic-homing torpedo launched by submarine. Unaffected by many forms of countermeasures. Modifications under way to allow it to be fired also from destroyers.

TORPEDO MARK 43 (Navy)
Lightweight acoustic antisubmarine torpedo launched from aircraft and surface ships. Length, 8 ft. Diameter, 10 in. Advanced version with greater speed and capable of greater depths. Rocket assisted. Operational. Clevite.

TORPEDO MARK 44 (Navy)
Acoustic antisubmarine torpedo launched from either aircraft or surface ships. Fall slowed by small parachute when launched from plane. On impact with water, parachute released, leaving torpedo free to seek its target. Capable of nuclear submarine speeds and depths. Also used as ASROC payload. Advanced version in R&D.

TORPEDO MARK 46 (Navy)
Lightweight acoustic torpedo. Capable of high speed, long range, and great depths. Solid propellant. Radar-homing device. In R&D. Aerojet-General.

TORY (AEC/Air Force)
Nuclear Reactor under development in PLUTO program. TORY II A-1 tests completed. TORY IIC, advanced version, will have more streamlined configuration. Will demonstrate feasibility of nuclear ramjet propulsion. In 45-sec test run, TORY II A-1 developed 10 thermal Mw, equal to about 2000-lb thrust. Reactor intended for a power level of 150 Mw. Livermore Radiation Lab.

TOW
Tube-launched, Optically tracked, Wire-guided antitank missile. Successor to ENTAC. Being evaluated by Army.

TR-1 (Navy)
Rocket designed to investigate the feasibility of utilizing steam rocket propulsion. Preliminary development aimed at production of low-cost meteorological rocket. Experiment, Inc.

TR-101
See HILO; PENNY.

TRAAC (Navy)
Transit Research And Attitude Control. Program to study stabilization techniques. Doorknob-shaped satellite launched with
TRANSIT IVB, 15 November 1961. Made successful separation
from TRANSIT. Also supplied data for particle-detection experiments of inner Van Allen radiation belts.

TRACE (Navy)
Checkout system developed for POLARIS missile by Nortronics.
System completely automates the checkout procedure, replacing
the go—no-go system. Reduces countdown time from 5 hr to
8 min.

TRADEWINDS III (Navy)
Transmission of radio signals over very long distances by
guiding the waves through "ducts" in the air formed between
dry and wet air layers.

TRADEX (Army)
Tracking RADar, EXperimental. Dual-band radar system operating in the UHF band and the L-band. Can use simultaneous
operation on two frequencies to work out gating techniques,
thus making it possible to collect rate data on re-entry bodies.
Part of Project PRESS. Located on Roi-Namur Island to study
incoming targets for ZEUS. RCA.

TRAILBLAZER (NASA)
Two-part program to improve methods of detecting bodies
entering the atmosphere at high velocities. Designation, ADVD-
58. A sheath of ionized particles, capable of being detected by
radar, surrounds these bodies, making them appear to be much
larger objects. TRAILBLAZER I, a six-stage test vehicle used
to launch 5-in. spheres. Four shots made. See METEOR. TRAIL-
BLAZER II, four-stage booster which launches 15-in. spheres.
Atlantic Research/Chance-Vought.

TRANSIT I (Navy/ARPA)
First shot of all-weather navigation satellite system. Designation, S-60-6A; S-60-10. Plan four or more in 450-mile orbit
around earth, with one to two launches per year. THOR-EPSILON
booster. System operational possibly in 1962. TRANSIT I fired
17 September 1959 but failed to orbit. Purpose of system: to
enable submarines and surface ships to fix position within less

than $\frac{1}{10}$ mile. TRANSIT IB: First navigational test satellite to achieve orbit. Fired 13 April 1960 into equatorial orbit from Cape Canaveral by THOR-EPSILON booster. Inclination to equator, 51°. Perigee, 239 miles; apogee, 472 miles. Weight in orbit, 1500 lb, including 265-lb satellite. TRANSIT IIA: First firing to orbit a double payload, TRANSIT IIA and GREB with same THOR-EPSILON booster on 22 June 1960 from Cape Canaveral. Satellites separated in space and still in orbit. TRANSIT IIA: Weight, 223 lb. Perigee, 389 miles; apogee, 658 miles. GREB (NRL): Weight, 42 lb. Perigee, 382 miles; apogee, 658 miles. Total weight in orbit, 1500 lb. TRANSIT IIIA: Satellite fired 30 November 1960 but failed to orbit. TRANSIT IIIB: Satellite fired by THOR-EPSILON booster with piggyback LOFTI 21 February 1961. Weight of TRANSIT IIIB, 250 lb; LOFTI, 54 lb. Perigee, 117 miles; apogee, 429 miles. TRANSIT dimensions same as previous TRANSITs. LOFTI diameter, 20 in. Satellites failed to separate but transmitted valuable data in spite of this. TRANSIT IVA: Double piggyback shot with GREB III and INJUN. THOR-EPSILON booster. Launched 29 June 1961 and orbit achieved. TRANSIT separated from the piggyback payloads, but they themselves failed to separate. Transmitting good data although some instrumentation obscured. GREB III purpose: to measure X-ray emanations from sun. INJUN instrumented to measure radiation intensity in the inner Van Allen belt. First use of nuclear energy in a space vehicle in TRANSIT IVA which carried a SNAP generator. Weights: TRANSIT, 175 lb; GREB, 55 lb; INJUN, 45 lb. Equatorial, polar orbit. Inclination to equator, 66.68°. Perigee, 550 miles; apogee, 629 miles. TRANSIT IVB: Successful launch 15 November 1961. Carried piggyback TRAAC to test stabilization device.

TRIBE (ARPA)
Study of family of advanced military space vehicles capable of performing designated military space missions, including guidance, stabilization, and control components, and including existing or planned boosters, e.g., AGENA, SATURN.

TRIDENT (France—Army)
Air-breathing surface-to-air missile. Solid propellant. Range, 24 plus nm. Speed, Mach 3.3. Ceiling, 130,000 ft. Radar-homing guidance. Ramjet engine. Operational. To be used for French space shots. Sud-Aviation.

TRIDENT (Navy-BuShips)
Project to develop equipment for ocean surveillance from shore.

Includes FISHBOWL, an active system; CAESAR, a passive system, and ARTEMIS, long-range sonar detection. In R&D.

TRIESTE
Hydrospace program involved in exploration of the 35,800-ft Challenger Deep in the Marianas Trench by means of a bathyscaphe. See also NEKTON II.

TRIGA (AEC)
Atomic reactor which can produce short pulses of intense nuclear energy. Purpose: to study radiation effects on electronic equipment. Operating prototype at Johns Hopkins Lab., San Diego, Calif. Contract to G.D.—General Atomic for reactor for Navy Diamond Fuze Labs, National Naval Medical Center, Bethseda, Md.

TRITON (Navy)
Surface-to-surface missile. Military designation, XSSM-N-2. Length, 47 ft. Gross weight, 20,000 lb. Cruising speed, Mach 3.5. Altitude, 80,000 ft. Range, 1500 miles. Cost, $24 million. Outgrowth of APL (JHU) BUMBLEBEE program. McDonnell, prime.

TROY 2A
Nuclear reactor built for PLUTO program to test nuclear-ramjet propulsion system in simulated flight. Marquardt.

TS-609A
See BLUE SCOUT JR.

TSX
Telephone Satellite eXperimental. Former designation for ATT TELSTAR communications satellite.

TU-122
See XM55.

TU-123
See XM56.

TULLIBEE (Navy)
First known true nuclear-powered antisubmarine submarine. Electric Boat.

TUMBLEWEED (AEC)
Project using the HAS (High Altitude Sampler) rocket, the NIKE-VIPER. Purpose: collection of radiation data. Operational. Sandia.

TV
Test Vehicle.

TV-1 (Navy)
Flight test vehicle composed of VIKING No. 14, with third stage of VANGUARD for a second stage.

TV-3
Navy VANGUARD engine.

TX-7-1
See M7A1.

TX-10
See XM10.

TX-12
See XM12.

TX-16-1
See M16E1.

TX-18, -1, -2
See XM18.

TX-18-4
See XM46.

TX-19; TX-19-1
See XM19; XM19E1.

TX-20
See XM20.

TX-30
See XM30.

TX-33
See XM33.

TX-52-1 (Navy)
Test vehicle for DOORKNOB. Solid propellant. Thiokol.

TX-58-1
See M58E1.

TX-62, -3
See XM43.

TX-83-4
See XM83-4.

TX-131-15
See XM51.

TX-135
Solid rocket engine used with NIKE-ZEUS.

TX-136
Military designation, XM6. A self-destroying, single-chamber booster intended to replace the TX-135 booster motor now used on the NIKE-ZEUS. Has glass-fiber and reinforced plastic case, reinforced plastic nozzle instead of steel case. Under development by Thiokol.

TX-148
See XM148.

TX-167-1
See XM26E3.

TYCHO (NASA)
Determination of the structure, the mass ejected from it, and other features of the ray system of the lunar crater Tycho. Decker Corp.

TYPE 2 (U.S.S.R.—Air Force)
Air-to-air and air-to-surface solid-propellant missile. Length, 15 ft. Weight, 1120 lb. Range, 5+ miles. Speed, Mach 2+. 550-lb HE warhead. Infrared guidance. Improved version of the M-100. Operational-development status.

TYPHON (Navy)
Air-to-air antiaircraft and antimissile missile. Nuclear or HE warhead. Two versions under development: Long Range TYPHON

(formerly SUPER TALOS) and Medium Range TYPHON (formerly SUPER TARTAR). Range, estimated, 100 nm. Speed, supersonic. May be used on hydrofoil destroyers. R&D. Westinghouse,
prime; Bendix, propulsion. Advanced guidance studies by Johns
Hopkins.

U

U.K. I, II (Great Britain/NASA)

Satellite series to be instrumented by United Kingdom and
lofted by NASA, THOR-DELTA booster, into orbit with 200-
mile perigee, 600-mile apogee. Six shots planned, one in 1962
and four in the 1963-64 time period. U.K. I to weigh 170 lb.
Diameter, 20 in. Primarily designed to study ionosphere, but
also to contain cosmic-ray experiment proposed by Imperial
College, London. Electron density in the ionosphere to be measured by experiment developed at University of Birmingham.
Two experiments developed by University College, London, to
obtain various parameters of the ionosphere by a detailed in-
terpretation of the voltage-current characteristics of an elec-
trode placed in the ionosphere. Emission of ultraviolet light
and soft X rays to be monitored by ionization chambers and
proportional X-ray counters. U.K. II experiments to measure
radio emissions from the galaxy at wavelengths too long to
penetrate earth's atmosphere; measure the vertical distribution
of ozone in the atmospheric region in which it is being formed;
obtain information on the electrical condition of the high atmosphere; measure the number and size of the fine dust particles
encountered by the satellite.

UNESCO

United Nations Educational, Scientific, and Cultural Organization.

UNICALL

An alpha-numerical input/output mechanism by which a computer can reply verbally to questions made from many remote
points. Developed by UNIVAC.

UNICOM

UNiversal Integrated COMmunications System. Switching and
terminal facilities developed by Western Electric for a reliable
world-wide command support communications network.

UNITAS I, II (U.S. Navy/South American navies)
Antisubmarine warfare training exercises held jointly by U.S.
Navy and navies of eight South American countries. I: 22 August
1960 to 14 December 1960. II: Held over a period of four months
beginning 7 August 1961.

UNITERM
A system of coordinate indexing used in libraries.

UPPER MANTLE (U.S.S.R.)
Soviet project to penetrate the earth's crust by drilling.

UPSTART (Navy/Army)
High-altitude target missile. Designed for vertical firing and
parachute recovery. Quite similar to POGO-HI but with greater
range and altitude. Speed, 4000 mph. Range, 170,000 ft. Test
launched March 1961. Aeronca.

URAL (U.S.S.R.)
A Soviet computer.

URANIA
Proposed space vehicle which would use nuclear-electrical
propulsion and follow a parabolic flight path. Capable of 10-man
space expeditions.

URBM
Ultimate Range Ballistic Missile.

USACE
U.S. Army Corps of Engineers.

USAREUR
United States ARmy, EURope.

USCONARC
U.S. CONtinental AiR Command.

USD
United States Drone.

USD-5
See OSPREY.

USM
Underwater-to-Surface Missile.

UVICON (Smithsonian Institution, Astrophys. Observatory)
    Electronic imaging tube sensitive to ultraviolet light, developed
    by Westinghouse Research Laboratories. To be used in Project
    CELESCOPE (OAO) to view universe as never before.

V

V-1 (German Army)
    Also termed FIESELER FI-103. Pulse-jet driven air-breathing
    vehicle. World's first operational long-range guided missile.
    Speed, 400 mph. Launched June 1944. See LOON.

V-2 (German Army)
    Liquid-propellant tactical missile developed at Peenemünde,
    Germany, under direction of Dr. Walter R. Dornberger and
    Dr. Wernher von Braun. Length, 46 ft. Launch weight, 28,335 lb.
    Ballistic (gyro and fin-stabilized) guidance. Speed, Mach 5.
    Range, 225 miles. Captured V-2 rockets were first in America
    capable of delivering large warhead at long range.

V-891
    See VIGILANT (Great Britain).

VALKYRIE
    Military designation, WS-110A. See B-70.

VANGUARD (Navy/NASA)
    Three-stage intermediary rocket booster. Navy designation,
    TV-3. Length, 72 ft. Weight, 11 tons. Liquid-solid propellant.
    Speed, 18,000 mph. Thrust, 27,000 lb. Inertial guidance. VAN-
    GUARD series' purpose: to determine exact shape of earth,
    measure magnetic field, and investigate solar X rays and space
    environment. VANGUARD I still transmitting. Program phased
    out. Cost, $125 million. Martin/NRL. See ORBITER.

VANGUARD I
    Also called MIGHTY MITE, SPACE METHUSELAH, and GRAPE-
    FRUIT. Satellite lofted 17 March 1958. Weight, 3.25 lb. Apogee,
    2455 miles; perigee, 404 miles. Estimated life in orbit, 200 to
    1000 years. Solar battery-powered radio still transmitting.
    Established precise pear-shape of the earth and determined
    the radiation pressure from the sun.

VANGUARD II
Satellite lofted 17 February 1959 to test weather-scan concept. Designation, SLV-4. Apogee, 2052 miles; perigee, 343 miles. Called the cloud-cover satellite. Still orbiting.

VANGUARD III
Satellite lofted 18 September 1959 to make radiation and space environment experiments. Weight, about 100 lb. Apogee, 2318 miles; perigee, 322 miles. Estimated life 50 to 150 years. Last shot of VANGUARD program.

VATE (Air Force)
Versatile Automatic Test Equipment. Purpose: computer control, specifically, standardization aimed at checkout and test for guidance in ATLAS, TITAN, SKYBOLT, HOUND DOG, MINUTEMAN. Study contracts to ARMA, Hughes, ITT, Sperry Gyroscope, and Lockheed Missiles and Space.

VAX (Navy)
Tactical aircraft, smaller and slower than the TFX but with same variable sweep wing, for support of Army and Marine ground troops. Should have definite STOL characteristics. Now seen to be an enlarged, improved version of Douglas A4D attack plane.

VB
Vertical Bomb.

VB-1
See AZON.

VB-3
See RAZON.

VB-10
See ROC-1.

VEDAS (Army/Air Force)
Space Vehicle/Missile Mapping and Geodesy System. Probably inactive.

VEGA (France)
Surface-to-air experimental vehicle. Ramjet-powered. Solid propellant. Length, 32.5 ft. Payload weight, up to 660 lb. Speed, Mach 4 to 5. Range, about 160 nm. Ceiling, 65,000 to 110,000 ft. Nord Aviation.

VEGA (NASA)
Liquid-propellant space booster. To have been used with Atlas. Capable of lofting 800-lb payload to Venus or 500-lb payload to Mars. Canceled December 1959 in favor of the AGENA B. Convair.

VELA (Air Force/AFBMD)
Aerospace project for control posts for detecting nuclear explosions. ATLAS-AGENA B booster. Three categories: VELA UNIFORM: purpose, detection of underground nuclear explosions. VELA SIERRA: purpose, ground-based detection of nuclear tests in space. VELA HOTEL: purpose, satellite-based detection of nuclear tests in space; two shots scheduled. Weight of payload, 300 to 400 lb. Program calls for participation by universities and other private organizations, in addition to various government agencies. Preliminary tunnel work done in the Tatum Salt Dome, and Briunsbury Salt Dome, Miss. G.E.-Tempo.

VELVET GLOVE (Canada)
Air-to-air guided missile project for use by Royal Canadian Air Force, 1953-54. Special Weapons, Canadair, Ltd. Montreal.

VENUS FLYTRAP
Sounding rocket lofted to collect and bring back space dust samples. Analysis of samples indicated that micrometeoroids are 100 times more dense than previously thought.

VERLORT
VERy LOng Range Tracking. Modified SCR-584 radar for use in MERCURY and DISCOVERER tracking stations. Reeves.

VERNISTATS
Precision potentiometers used with TALOS, TITAN, SUBROC, POLARIS.

VERONIQUE (France)
Sounding rocket capable of lofting 116-lb payload to 140-mile altitude. Laboratoires de Recherches Balistiques et Aérodynamiques.

VERTISTAT
A mechanical device to maintain attitude of satellites with respect to planets. Unit is based on the "gravity gradient erection principle." G.D.—Astronautics.

VEWS (Air Force)
Very Early Warning System. Includes space vehicle, tropospheric scatter links, tracking stations, and ground network. Study contract to Douglas, 1960.

VFR
Visual Flight Rules. Rules used by pilots in navigating and landing an airplane when visibility is satisfactory.

VIGILANT (Great Britain)
Surface-to-surface antitank missile. Solid propellant. Speed, 350 mph. Range, 1700 yd Wire guidance. Two types HE warhead. Operational. Private venture by Vickers-Armstrong.

VIKING (Navy)
Formerly NEPTUNE. Military designation, RTV-N-12. Fourteen liquid-propellant rockets (no two alike) built and fired 1949-1957. Successor to V-2 models ranging in length from 42 to 48.5 ft, carrying payloads from 464 to 887 lb. Weight loaded, from 9650 to 15,035 lb. Speed, 1800 to 4300 mph. Held world record for single-stage rockets with an altitude of 158.4 miles, until surpassed by AEROBEE. Martin.

VINITI (U.S.S.R.)
Soviet information service which abstracts about half a million technical articles each year.

VIPER (NASA/Air Force)
Experimental missile. Solid propellant. Used in clusters to propel sled vehicles at USAF test tracks. Operation, 1960. Lockheed Propulsion (formerly Grand Central Rocket).

VIPER-ARROW
Small sounding rocket capable of lofting a 12- to 20-lb payload to 100-mile altitude. Two-stage vehicle. Length, 17 ft. Solid propellant. Composed of Grand Central VIPER for first stage, ARROW for second stage. Low cost. Developed with DEACON-ARROW.

VIPER FALCON
Sounding rocket capable of lofting 20-lb payload to 75-mile altitude. Two flown and three ordered for Navy. Zimney.

VIPERSCAN (Navy)
VIPER rocket with scanner payload lofted to 250,000-ft altitude.

Scanner then ejected and triggered to send its pictures back to earth. NOTS.

VIRTUE
Philco reliability assurance program. Uses 100% automatic life test equipment. The company goal is a composite transistor failure rate of 0.001%/1000 hr of operation. Cost of program, $5 million.

VISILOG
A bionic machine following the principles of the human eye. To be capable of collision avoidance, terminal guidance for soft landings, and operation of vehicles over unusual terrain. Optical means used to measure slant range and relative distance. G.E.— Light Mil. Elect. Dept.

VN-101
Vehicle proposed to demonstrate feasibility of low-cost drone. Was to have gold-plated rocket motor. Lockheed.

VOICE COMMANDER
Portable, two-way radio. One-watt transmitter radio frequency output for high-band military frequencies. Weight, 4 lb. Dimensions, 9.5 × 5.3 × 1.7 in. G.E.

VOIS (NASA)
Visual Observation Instrumentation Subsystem. Photographic equipment probably to be mounted on SURVEYOR lunar orbiter, to map moon to 100-m resolution. Three VOIS systems proposed to JPL by Eastman Kodak, Fairchild Camera, and RCA in early 1961; one will be selected.

VOODOO
Air Force fighter bomber, developed by McDonnell. Designation, F-101B.

VORTAC
High-frequency radio range system used for terminal landing of aircraft.

VOSTOK I (U.S.S.R.)
Also called SPUTNIK XI; SPACECRAFT VI; KOSMICHESKIJ KORABL-SPUTNIK VI. Experimental satellite launched 12 April 1961 from Baykonur, U.S.S.R. Spacecraft weight (less final stage), 10,415 plus lb. Inclination, 65°57'. Perigee, 112.4 miles;

apogee, 203 miles. Multistage rocket booster reportedly with six engines having total thrust of 20 million hp at launch. Carried Major Yuri Gagarin, first human being with successful recovery during first orbit. Recovered in Saratov area 12 April 1961 after one earth orbit.

VOSTOK II (U.S.S.R.)

Also called SPUTNIK XII; SPACECRAFT VII; KOSMICHESKIJ KORABL-SPUTNIK VII. Second manned orbital shot. Launched at Baykonur, U.S.S.R., 6 August 1961. Spacecraft weight (less final stage), 10,431+ lb. Estimated to be 12 to 14 ft in diameter and 20 ft long. Inclination to equator, 64°56'. Perigee, 110.6 miles; apogee, 159.7 miles. Multistage rocket booster, reportedly with six engines with total thrust of 20 million hp at launch. Manned by Major Gherman Titov. Recovered in Saratov area after 17.5 earth orbits.

VOYAGER (NASA)

Spacecraft designed for Venus and Mars and interplanetary space research. Weight, 10,000 lb. To be capable of orbiting these planets, or ejecting impact loads, or both. SATURN booster to be used for VOYAGER I, II. Seven flights planned to 1971; five flights in the 1971-75 time period.

VOYAGER I

Development flight to Venus scheduled for 1965. Capsule with instrumentation similar to that of MARINER to be ejected for atmospheric entry and landing on surface. Data to be stored, relayed, or received directly on earth.

VOYAGER II

Spacecraft scheduled to be sent into interplanetary space out of the plane of the ecliptic.

VOYAGER III

Spacecraft planned for a Mercury flyby to study radiation belt and magnetic field. Formerly to be boosted by SATURN C-2 before this vehicle development was dropped.

VOYAGER IV

Spacecraft in Jupiter flyby to study radiation belt and magnetic field. Also was to have been lofted by SATURN C-2.

V/STOL

Vertical/Short Take-Off and Landing.

VTOL
   Vertical Take-Off and Landing.

VZ-10 (Army)
   HUMMINGBIRD VTOL test bed. Lockheed-California.

W

WAC CORPORAL (Army)
   Rocket developed by JPL under ORDCIT program. Combined
   with V-2 rocket in Project BUMPER after World War II. Established world record of two-stage rockets 24 February 1949,
   with 250-mile altitude. Length, 16 ft. Weight, 665 lb. Ceiling,
   43 miles. First rocket to use red fuming nitric acid and gasoline as propellant. Capable of lofting 25-lb payload. Succeeded
   by AEROBEE.

WADD
   Wright Air Development Division. Together with the Aeronautical
   Systems Center of AMC, now forms the Aeronautical Systems
   Division of AFSC.

WAGMIGHT (Navy)
   Inflatable vehicle capable of flying as manned aircraft, missile,
   or drone. Has vertical take-off capability from "accumulator"
   compressed air principle. Goodyear.

WAG TAIL (Air Force)
   Air-to-air and air-to-surface missile. Liquid propellant for
   air-to-air version. Solid propellant for air-to-surface model.
   Range, over 25 miles. Low-altitude launch. Engine (solid):
   XM43. Being developed to follow terrain at low levels. In R&D.
   Minneapolis-Honeywell, prime.

WALLOPS STATION (NASA)
   Launching site which has the principal responsibility of firing
   sounding rockets for upper-atmosphere research and for launching SCOUT.

WALNUT
   Electronic retrieval system. Capable of locating in 5 sec any
   page of thousands of documents on file in an information center,
   transferring a reduced image of the data to a card, and en-
   larging it on a viewing screen. System large enough to hold
   300,000 average-sized books. IBM.

WASP (Army/Navy)
Window Atmospheric Sounding Projectile. Solid-propellant research rocket to gather data from 20- to 30-mile altitudes. Length, nearly 9 ft. Weight, 24 lb, with Type-1 booster; 29 lb, with Type-2 booster. Altitude, 21 miles, with Type-1 booster; 29 miles, with Type-2 booster. Marquardt—Cooper.

WASSERFALL (Germany)
World War II missile. See C-2 (U.S.S.R.).

WEAPON ABLE
See WEAPON ALFA.

WEAPON ALFA (Navy)
Antisubmarine, surface-to-underwater missile launched from turret guns. Weight, 500 lb. Length, about 8.5 ft. Solid propellant. Free-flight guidance. Operational since 1952. Deployed on destroyer escorts and 931-class frigates. Developed by NOTS. Also called ALFA, ALPHA, ABLE, WEAPON ALPHA, WEAPON ABLE. Navy, prime; Avco, airframe.

WEAPON ALPHA
See WEAPON ALFA.

WEDGE
WEapon Development Glide Entry. Study of bomber-reconnaissance vehicle version of DYNA-SOAR. Company-financed project by Boeing. See BOSS-WEDGE.

WEEVIL (Army)
Rocket used as part of Army helicopter armament.

WEST FORD
Also called NEEDLES. A project calling for launch into 2100-mile orbit of about 350 million hair-like copper wires about 0.7 in. long, for use as dipoles in a world-wide communications experiment. The "needles" are packed in a cylinder 20×6 in. Weight of total package, 100 lb. Weight of dipoles, 75 lb. Project protested by many of world's astronomers on grounds that the needles would impair astronomical observations. WEST FORD was launched successfully 21 October 1961 from Point Arguello, Calif., as companion payload to MIDAS IV, launched by ATLAS-AGENA B. Result of experiment: needles did not disperse as planned. Project conceived by two Thompson Ramo Wooldridge scientists and developed at M.I.T. Lincoln Lab.

WHIP
Wideband High Intercept Probability receiver. Radically new electronic countermeasures system developed by Hallicrafters to provide a low-cost lightweight system for reconnaissance satellites. Capable of identifying multiple radio frequencies to within 1% accuracy. Receiver holds its accuracy over an octave of bandwidth and has a 100% intercept probability. Stanford University (Electronics Lab) scientists made original suggestion for WHIP system.

WHITE ALICE
Communication system in the Far North, consisting of microwave and FPTS (forward propagation tropospheric scatter) systems. Included 33 stations connected by ultra high-frequency tropospheric scatter radio and microwave trunk routes for more than 3000 miles across Alaska. Systems maintained entirely by contract. Investment, about $135 million.

WHITE LANCE (Air Force)
Air-to-surface missile with nuclear warhead. TV-guided. Outgrowth of BULLPUP. Discontinued. Martin.

WILLOW (Army)
No details available. Chrysler, prime.

WINDSONDE
Air-launched aluminum and magnesium missile system. Length of missile, 27 in. Purpose: to measure wind direction and velocity at a wide range of altitude. Allied Research Associates (Boeing subsidiary).

WIZARD (Air Force)
Antimissile missile system. First stage, range of 1000 miles. Second stage effective at altitudes up to 500 miles. Discontinued January 1958. Convair/RCA.

WOLF (NASA)
FLOW spelled backwards. Reverse flow solid rocket. 500-lb prototype built by Grand Central Rocket (Lockheed Propulsion). No specific application planned.

WOO (NASA)
Western Operations Office, located at Santa Monica, Calif.

WOOFUS (Navy)
Sounding rocket. Discontinued.

WOSAC
WOrld-wide Synchronization of Atomic Clocks.

WS
Weapon System.

WS-101A (Air Force)
High-altitude SAC bomber system (B-52). Fairchild.

WS-102A/L
See DUCK.

WS-103A
See SNARK.

WS-107A; WS-107A2
See TITAN.

WS-107A-1
See ATLAS.

WS-110A
See B-70.

WS-117L
See MIDAS.

WS-121B
See ARM.

WS-123A
See GOOSE.

WS-125A (Air Force)
Nuclear-powered strategic bomber system. Convair, G.E.

WS-126A
See BDM, HAWK.

WS-131B
See HOUND DOG.

WS-132A (Air Force)
Bomber defense missile for the B-70. Under development by
G.E./McDonnell and Republic/Westinghouse teams when work
was ordered suspended in November 1956.

WS-133A
　　See MINUTEMAN.

WS-138A
　　See SKYBOLT.

WS-169A
　　See BOLD ORION, SKYBOLT.

WS-199B, C
　　See SKYBOLT.

WS-200A
　　See BOMARC.

WS-202A
　　See RAPIER.

WS-212L
　　See BADGE.

WS-214L (Air Force)
　　Airborne early warning and communication system. G.E.

WS-216L
　　See SAGE.

WS-221A
　　See SIDEWINDER.

WS-222A
　　See WIZARD.

WS-238; WS-T238
　　See BOLT.

WS-239A
　　See MIDAS.

WS-299A (Air Force)
　　Air defense system studies.

WS-309A
　　See MACE.

WS-315A
See THOR.

WS-321A
See BULLPUP.

WS-323A
Tactical strike-reconnaissance system.

WS-398B
Theater ballistic missile TMX.

WS-399A
Tactical system studies.

WS-399B
Zero-length launch study.

WS-404
See MISSILEER.

WS-425L (Air Force)
NORAD combat operation center.

WS-426/2 (Navy)
Navy model of OQ-19. See KD2R-5.

WS-433L
Weather reconnaissance system.

WS-438L (Air Force)
Intelligence data-handling system.

WS-460L
Air Weather Reconnaissance Aircraft support system. Canceled after expenditure of $13.3 million.

WS-462L; WS-462/2 (Air Force)
See OQ-19.

WS-464L
See DYNA-SOAR.

WS-465L (Air Force)
Strategic Air Command Control system.

WS-466L (Air Force)
Electromagnetic intelligence system. Provides advanced tactical and strategic warning of imminent enemy operations.

WS-473L
See GLOBE QUICK.

WS-477L
See NUDETS.

WS-480L (Air Force)
Expansion of existing global Air Force communications.

WS-496L
See SPACETRACK.

WS-601L
Advanced Aerospace Offensive System. Under consideration by Air Force, 1961.

WS-602L
Advanced Aerospace Defensive System. Under consideration by Air Force, 1961.

WS-603L
Tactical Forces System. Under consideration by Air Force, 1961.

WS-609A
See BLUE SCOUT. Also called Air Force SCOUT.

WS-638A
See SKYBOLT.

WS-649L (Air Force)
Space combat weapon system.

WS-655A
See SLAM.

WS-810
See COBRA (Marines).

WSEM
Weapons System Evaluator Missile. A FALCON with clipped wings used for evaluation of missiles and for training. Hughes.

WSR-57
Weather detection radar. Raytheon.

X

X-1
Experimental nonflying rocket engine. Rocketdyne.

X-1, -1A, -1B, -1E (Air Force)
Straight-winged research aircraft, rocket-powered. Pioneered in world's first supersonic flight. Maximum speed of X-1: 967 mph; X-1A, Mach 2.5. X-1E reached greater speeds because of its lighter airframe. Bell Aircraft.

X-2 (Air Force)
Rocket-powered research plane. Set world's altitude record for manned flight in September 1956. Reached nearly 24-mile altitude. Also first aircraft to explore aerodynamic heating. Bell Aircraft.

X-3 (Air Force/Navy/NASA)
Jet-propelled aircraft. Purpose: to test design of an aircraft suitable for sustained flight at very high speeds. Douglas.

X-4 (Air Force/NACA)
Research plane. First U.S. production jet engine. Plane powered by two turbo-jets. Northrop.

X-5 (Air Force/NACA)
Jet-powered aircraft developed over period of 3 years to study transonic flight. First plane to feature variable wing in order to reduce compressibility at sonic speeds.

X-7; X-7A, -7B (Air Force)
Also known as METHUSELAH. An air-launched ramjet drone used for testing supersonic ramjet engines. Boosted to flight speed by solid-propellant rocket engine. Length, about 32 ft. Wing span, about 12 ft. Total fueled weight, 3400 lb. Recovery, by parachute and spike. Flights made from 1951 to 1960 with high record of recoveries. Supported BOMARC program. X-7B was version designed to test guidance. XQ-5 (KINGFISHER) was target drone version. Lockheed Missiles and Space; Marquardt, propulsion (ramjet); Thiokol, propulsion (solid).

X-10 (Air Force)
Guidance and control test bed for NAVAHO second stage. Canceled. NAA.

X-15 (Air Force/NASA/Navy)
Advanced experimental rocket-powered airplane, launched from a B-52 at 45,000 ft. Length, 50 ft. Span, 22 ft. Wing area, 200 ft^2. Gross weight, 31,275 lb. Fuel weight, 19,000 lb. Propulsion: First system, a 16,000-lb-thrust liquid engine consisting of two four-barrel chambers (Thiokol XLR-11); later, a 57,000-lb-thrust liquid rocket (Thiokol XLR-99) which was finally increased to 60,000-lb thrust at sea level or 70,000-lb thrust at peak altitudes. Program started in 1952. Contract awarded to NAA in December 1955. First X-15 delivered to Air Force in October 1958. Initial powered flight, September 1959. Twenty-nine powered flights made with XLR-11 engines; 15 flights with XLR-99 engines; and one unpowered flight. Performance data as of December 1961: Maximum altitude, 217,000 ft. Mach 6.04; 4093 mph. Maximum skin temperature, 1100° F. Flights scheduled for 1963. X-15 can be considered first manned aerodynamic space vehicle. To provide basic information for DYNA-SOAR and APOLLO. Cost of program: $172 million (AF); $45 million (NASA); $7 million (Navy).

X-17 (Air Force/NASA)
Three-stage hypersonic re-entry test missile used to test re-entry of nose cones. A pioneer program aimed to solve the re-entry problem. Length, 40.6 ft. Spin- and fin-stabilized. First stage, XM20 SERGEANT rocket; second stage, three XM19 rockets; third stage, XM19E1 rocket. First stage fired vehicle to about 100-mile altitude; other two stages fired on down-leg of trajectory to increase speed to more than Mach 10. Thirty-six flights were made (plus three nuclear ARGUS shots), 1955 to 1958, with high reliability. X-17 program paved way for other programs, e.g., POLARIS, SCOUT. Lockheed Missiles and Space; Thiokol, propulsion.

X21B12 (Navy)
Submarine simulator. Clevite.

X-213
Solid-propellant sustainer used for TERRIER. Also booster for TERRIER and booster and third stage for NIKE-AJAX. Hercules Powder.

X-216
Solid-propellant engine used for JAVELIN second stage. Hercules Powder.

X-220
Engine for one stage of DEACON. Thrust, 5700 lb. Solid propellant. Hercules Powder.

X-223
Solid-propellant sustainer for BULLPUP. Hercules Powder.

X-226
Solid-propellant booster for SNARK. Hercules Powder.

X-235
Solid-propellant engine for LITTLE JOHN. Hercules Powder.

X-244
Solid-propellant engine used for HONEST JOHN. Also first-stage engine for JAVELIN. Hercules Powder.

X-248
Rocket developed for third stage of VANGUARD, DELTA, and ABLE vehicles. Now known as ALTAIR. First fully developed rocket to be constructed of lightweight plastic throughout.

X-250A2
Solid-propellant engine for POLARIS second stage. Hercules Powder.

X-251
Solid booster for TALOS. Hercules Powder.

X-254
See ANTARES. Also used as third stage of SCOUT and BLUE SCOUT. Thrust, 13,800 lb. Solid propellant. Hercules Powder.

X-405H
VEGA engine. 35,000-lb thrust. G.E.

XAAM-A-2
See FALCON.

XAAM-N-4
See ORIOLE.

XAAM-N-10
  See EAGLE.

XAE (Army)
  Drone under development. Aerojet-General.

XASM-N-2
  See PETREL.

XASM-N-4
  See DOVE.

XASM-N-5
  See GORGON.

XASM-N-6
  See OMAR.

XASM-N-8
  See CORVUS.

XASM-N-9
  See RAVEN.

XASR-SC-1, -2
  Military designation for Army models of AEROBEE-HI.

XAUM-N-2
  See PETREL.

XAUM-N-6
  See PUFFIN.

XGAM-87A
  See SKYBOLT.

XIM-99
  See BOMARC.

XKD2B-1
  See KD2B-1.

XKD2R1
  See KD2R1.

XKD4R-1; XKDR-1
Navy model of RP-70.

XKDT-1
See TEAL.

XLR-11 (Air Force)
Interim engine used for the X-15 research aircraft. Two XLR-11 engines, each with 8000-lb thrust, used. Thiokol. See XLR-99.

XLR 70 NA-9 MB-3
Liquid booster, 150,000-lb thrust, for THOR. Rocketdyne.

XLR81-BA-3, -5, -7, -9
Upper stage engine for ATLAS/THOR-AGENA. Liquid propellant. Thrust for XLR81-BA-9, 16,000 lb. Bell Aerosystems.

XLR-99 (Air Force)
Final engine used for X-15. Thrust, 57,000 lb. Thiokol.

XLR-99-RM (Air Force)
Engine for X-15. Reaction Motors.

XLR-113-AF
Liquid engine under development by Aerojet-General.

XLR-115-F; XLR-115-P-1; LR-115 (Air Force)
Liquid-propellant engine for CENTAUR second stage. Thrust, 15,000 lb. Now designated RL-10. Pratt & Whitney.

XLR-119
Hydrogen-oxygen engine. Uprated version of the XLR-115 (RL-10). Designed to produce 17,500-lb thrust. Pratt & Whitney.

XM
Research missile.

XM3E1
See HAWK.

XM4E1
See CORPORAL.

XM6
See TX-136.

XM6E3 (Army)
See NIKE-HERCULES.

XM10
Solid booster used with LACROSSE. Thiokol.

XM12
Booster with 63,000-lb thrust, used with SERGEANT.

XM18; XM18E1; XM18E3
Solid-propellant sustainer engines used with FALCON series.
Thiokol.

XM19E1
See RECRUIT.

XM19; XM19E1
Also TX-19; TX-19-1. Solid-propellant sustainer engine used
with ARGUS and FARSIDE. Thiokol.

XM20
Also TX-20. Solid-propellant engine used with re-entry vehicles.
Thrust, 49,940 lb. Thiokol.

XM21
See FIREBEE (Army).

XM26E3
Also TX-167-1. Solid rocket engine used with LITTLE JOHN.
Thiokol.

XM28/29 (Army)
Nuclear mortar. Two versions: one portable by personnel for
direct short-range; the other heavier, with longer range. Ford-
Aeronutronic.

XM29
Solid rocket motor used with GENIE. Thiokol.

XM30
Also TX-30. Solid-propellant sustainer engine used with NIKE-
HERCULES. Thiokol.

XM33
Also TX-33. Solid-propellant booster used with POLARIS.
Thiokol.

XM36
Solid-propellant sustainer engine for POLARIS. Thiokol.

XM38
Solid-propellant rocket motor used for GOOSE. Thiokol.

XM43
Solid-propellant engine for original air-to-surface version of WAGTAIL. Thiokol.

XM45 (NASA)
Capsule escape rockets. ABMA/Thiokol.

XM46
Solid-propellant sustainer engine used in FALCON series. Thiokol.

XM47
See LITTLE JOHN.

XM50
See HONEST JOHN.

XM51
Also TX-131-15. Solid-propellant rocket motor for BOMARC booster.

XM53
Solid booster used for SERGEANT. Sperry.

XM55
Also TU-122. Solid rocket motor for MINUTEMAN first stage.

XM56
Also TU-123. Solid rocket motor for MINUTEMAN second stage.

XM72
See LAW.

XM83-4
Also TX-83-4. Solid rocket motor (steering) used with JUPITER. Thiokol.

XM148
See JUPITER; TX-148.

XM388
  See DAVY CROCKETT.

XM474
  See PERSHING.

XQ-2B, -2C
  See FIREBEE (Air Force)

XQ-5
  See KINGFISHER.

XSAM-A-7
  See NIKE-AJAX.

XSAM-A-25
  See NIKE-HERCULES.

XSAM-N-6
  See TALOS.

XSM-64
  See RISE.

XSM-65
  See ATLAS.

XSM-80
  See MINUTEMAN.

XSSM-A-1
  See MATADOR.

XSSM-A-2
  See NAVAHO.

XSSM-A-13
  See HERMES (A-2).

XSSM-A-14
  See REDSTONE.

XSSM-A-16
  See HERMES.

XSSM-A-23
  See DART.

XSSM-N-2
See TRITON.

XSSM-N-6
See RIGEL.

Y

Y (NASA)
Liquid hydrogen engine anticipated for use in upper stages of SATURN C-5. Thrust, 2 million lb. To be developed by either Aerojet-General or Rocketdyne.

YM
Prototype missile designation.

YO-YO (ARPA)
Picture-taking, reconnaissance satelloid to be launched from high-speed fighters. Would make one orbit and be recovered at sea. One test at China Lake, Calif. Navy proposal. Now a project. May be transferred to Air Force.

Z

ZEL
Former name for REGULUS II. See M-34, MEGABOOM.

ZEMAR (Army-Ordnance Corps)
NIKE-ZEUS anti-ICBM system. Second generation. Under study by Army Ordnance.

ZERO GRAVITY BELT
Individual propulsion system developed by Bell Aerosystems for use by a free-floating man orbiting in space. Purpose: to permit inspection and maintenance on exterior of orbiting space vehicles. Propellant used by prototype was super-dried high-pressure nitrogen. Operational model may use H_2O_2 which was used in the Army Rocket Belt, also developed by Bell.

ZEUS
See NIKE-ZEUS; PINCUSHION.

ZUNI (Navy)
Solid-propellant missile. Length, 9 ft. Weight, 107 lb. Slant-range air-to-air missions, 3000 to 10,000 ft. Ground attack,

3000 to 25,000 ft. Speed, Mach 2+. Infrared homing guidance. Operational. Cost, $400 each. NOTS, prime; Hunter Douglas Div. of Bridgeport Brass.